Industry 4.0

Interdisciplinary Thought of the 21st Century

Management, Economics and Law

Series Editors
Elena G. Popkova and Artem I. Krivtsov

Volume 4

Industry 4.0

Implications for Management, Economics and Law

Edited by
Marina L. Alpidovskaya, Ludmila A. Karaseva,
David I. Mamagulashvili, Aleksei V. Bogoviz
and Artem I. Krivtsov

DE GRUYTER

ISBN 978-3-11-065065-5
e-ISBN (PDF) 978-3-11-065448-6
e-ISBN (EPUB) 978-3-11-065074-7
ISSN 2626-7063

Library of Congress Control Number: 2020951679

Bibliographic information published by the Deutsche Nationalbibliothek
The Deutsche Nationalbibliothek lists this publication in the Deutsche Nationalbibliografie;
detailed bibliographic data are available on the Internet at http://dnb.dnb.de.

© 2021 Walter de Gruyter GmbH, Berlin/Boston
Cover image: janeb13/pixabay.com
Typesetting: Integra Software Services Pvt. Ltd.
Printing and binding: CPI books GmbH, Leck

www.degruyter.com

Introduction: Challenges for Management, Economics and Law during a Transition to Industry 4.0

Industry 4.0 is a new digital and technological mode, a transition to which is performed by more and more developed and developing countries around the world. While initially Industry 4.0 was an investment project, now it is the basis of economy's modernization and is adopted as a strategic priority in the national programs of economic systems' development. Industry 4.0 envisages intensive development of high technologies, hi-tech production, and exports.

The international recognition of Industry 4.0 is proved by the fact that the World Economic Forum renamed its index to the Global Competitiveness Index 4.0, emphasizing the fact that global competitiveness is achieved by means of success in formation of Industry 4.0. In 2019, IMD added new indices to the World Digital Competitiveness Report, which reflect the development of Industry 4.0: "Robots in Education and R&D", "Use of big data and analytics", and "World robots distribution". To study robotization as one of the directions of Industry 4.0 – robotization of industry – a special international organization, International Federation of Robotics, was created. This organization collects and analyzes statistics on robotization from around the world and compiles international rankings.

In 2020's, Industry 4.0 ceases to be a theoretical concept – now it is widely applied in practice. The leading technologies of Industry 4.0 in the spheres of robotization, artificial intelligence, the Internet of Things, blockchain, cloud technologies, and others went beyond research labs and found their reflection in the economic practice of digital companies, information society, and e-government.

However, this process has not been studied sufficiently in scientific literature, since Industry 4.0 has been considered only from the positions of technologies and algorithms of their implementation into production based on creation of cyber-physical systems. New economic practices, social interactions, and regulatory mechanisms, which are used during the practical application of the technologies of Industry 4.0, are not studied sufficiently in the existing literature, which is a research gap. This book aims at filling this gap by considering challenges for management, economics, and law during a transition to Industry 4.0 and offering perspective answers to them. Originality and uniqueness of this book are derived from its following advantages.

Firstly, the book presents a systemic multidisciplinary view at Industry 4.0 from the positions of management, economics, and law. This allows determining the role of Industry 4.0 for economic growth and global competitiveness of economy and finding the specifics of management and state regulation of Industry 4.0. Due to this, the book is a guide for practical implementation of national strategies of economic systems' digital modernization based on transition to Industry 4.0.

Secondly, the book contains not only theoretical provisions, which form a comprehensive scientific vision of Industry 4.0 as an economic & legal and managerial concept, but also applied solutions in the sphere of management, economics, and law. This is what makes the book useful for socio-economic systems of different levels – from companies in Industry 4.0 to governments that regulate Industry 4.0. Due to this, the book will be interesting for a wide audience, which stretches far beyond the academic community, including company managers and state regulators of Industry 4.0.

Thirdly, the book reflects the leading international experience of transition to Industry 4.0. This allows forming a complex vision of Industry 4.0 and developing its universal scientific concept and universal applied solutions for development of Industry 4.0 in the sphere of management, economics, and law. Considering international experience is especially useful for determination of differences in the regulatory support for Industry 4.0 and the perspectives of its improvement and unification.

Fourthly, the book presents a new view at Industry 4.0, which is treated not as impersonal technical system but as a social institution – flexible and complex – which envisages social interactions and depends on public opinion. This sheds light on a previously hidden – social – aspect of transition to Industry 4.0. The book determines and considers current socio-economic, managerial, and legal problems of transition to Industry 4.0 and offers recommendations for solving them in the sphere of management, economics, and law.

The book consists of four parts, which consistently describe the essence and specifics of Industry 4.0 from the positions of management, economics, and law. Part I considers innovations and economic growth in Industry 4.0; reflects identifying fast-growing companies to differentiate government support measures; considers economy without finance – situation in R&D; studies innovative activity in post-pandemic conditions; and outlines the challenges of digitalization for economic relations of tourism industry.

Part II is devoted to finances, corporate accounting, and management in Industry 4.0. The chapters here dwell on problems of balancing of Russian foreign loans market, the problem of monitoring the effectiveness of projects in the process of their implementation, the use of digital technologies in the implementation of internal control in the management accounting system, and tactical and technical solutions for intra-operative critical situations.

Part III considers sustainable development in Industry 4.0: digital modernization in enterprises of agricultural and water management industry, the role and value of production accounting in providing the company's microeconomic stability, sharing economy and the influence of digital technologies on economic reality, and the impact of the Fourth Industrial Revolution on the socio-economic development of the world economy.

Part IV is devoted to high technologies and state management in Industry 4.0. The chapters of Part IV study economics of Industry 4.0 in the political economy paradigm, features of company's organizational culture, the interests of the state and business in stakeholder management, and methodological approaches to tender procedures in corporate procurement management in companies.

Marina L. Alpidovskaya, Ludmila A. Karaseva, David I. Mamagulashvili, Aleksei V. Bogoviz and Artem I. Krivtsov

Contents

Introduction: Challenges for Management, Economics and Law during a Transition to Industry 4.0 —— V

Part I: Innovations and Economic Growth in Industry 4.0

Elena A. Fomina, Julia V. Khodkovskaya, Ilya I. Beloliptsev and Denis V. Chuvilin
1 Challenges in Identifying Fast-Growing Companies to Differentiate Government Support Measures —— 3

Konstantin N. Lebedev, Yuliya I. Budovich and Imomnazar E. Tursunov
2 Economy without Finance: Situation in R&D —— 13

Gulnora Sh. Karabaeva, Rano R. Nazarova and Gulchexra N. Nigmatullayeva
3 Innovative Activity in Post-Pandemic Conditions —— 19

Evgeniya K. Karpunina, Zulay K. Tavbulatova, Yuri V. Kuznetsov, Naida D. Dzhabrailova and Olga A. Anichkina
4 The Challenges of Digitalization for Economic Relations of Tourism Industry Subjects —— 31

Part II: Finances, Corporate Accounting and Management in Industry 4.0

Irina G. Sergeeva, Irina E. Zuber, Vera D. Nikiforova and Alexander A. Nikiforov
5 Russian Foreign Loans Market: Problems of Balancing —— 47

Elena G. Patrusheva, Elena I. Lifanova and Anna V. Raikhlina
6 The Problem of Monitoring the Effectiveness of Projects in the Process of their Implementation —— 55

Nodira B. Abdusalomova and Zarina U. Tashkenbaeva
7 The Use of Digital Technologies in the Implementation of Internal Control in the Management Accounting System —— 61

Ismatilla T. Ydyrysov, Keneshbek B. Yrysov, Emir Z. Tuibaev,
Zhenishbek A. Kochkonbaev and Oyatilla A. Umurzakov
8   Tactical and Technical Solutions for Intraoperative Critical Situations —— 69

Part III: Sustainable Development in Industry 4.0

Shakhlo T. Ergasheva and Rano A. Mannapova
9   Digital Modernization in Enterprises of Agricultural and Water Management Industry —— 77

Rasul O. Kholbekov and Feruzakhon R. Kholbekova
10   Role and Value of Production Accounting in Providing the Company's Microeconomic Stability —— 85

Valery V. Gusev, Gamzat U. Magomedbekov, Gulnaz F. Galieva, Marina A. Gundorova and Zhanna A. Shadrina
11   Sharing Economy: How Digital Technologies Have Changed Economic Reality —— 95

Aleksei Tebekin, Ekaterina Bogoeva, Andrei Zakharov and Dmitrii Lazarev
12   The Impact of the Fourth Industrial Revolution on the Socio-Economic Development of the World Economy —— 107

Part IV: High Technologies and State Management in Industry 4.0

Kamila V. Kudryavtseva, Moisey A. Skliar, Lidiya R. Vakhitova and Natalya A. Shapiro
13   Economics of Industry 4.0 in the Political Economy Paradigm —— 115

Marianna S. Santalova, Igor L. Surat, Dariko K. Balakhanova, Irina V. Soklakova and Vladimir I. Surat
14   Features of the Company's Organizational Culture —— 123

Elena A. Fomina, Julia V. Khodkovskaya, Ilvir I. Fazrakhmanov
and Ekaterina E. Barkova
15 Harmonizing the Interests of the State and Business in Stakeholder Management —— 133

Lubov I. Vanchukhina, Tatyana B. Leybert, Elvira A. Khalikova,
Ilnara R. Khanafieva and Giedrius Ciras
16 Methodological Approaches to Tender Procedures in Corporate Procurement Management in Companies —— 141

Conclusion: Future Perspectives of Industry 4.0 —— 151

List of Figures —— 153

List of Tables —— 155

Index —— 157

Part I: **Innovations and Economic Growth in Industry 4.0**

Elena A. Fomina, Julia V. Khodkovskaya, Ilya I. Beloliptsev
and Denis V. Chuvilin

1 Challenges in Identifying Fast-Growing Companies to Differentiate Government Support Measures

Introduction

Interest in rapidly growing companies (RGC) – "gazelles" – in the world and Russia is supported to a large extent due to the results of assessing their impact on indicators such as job creation and participation in income growth in the economy of a particular region or country as a whole. Since 2000, many researchers have confirmed that the RGC effect persists if it does not increase. Therefore, in the UK, 6% of all firms with 10 or more employees created 54% of jobs in 2005–2008. In Sweden, this ratio is 6–42%, in Finland 5–90% (in 2003–2006) (Goswami, 2019). In the United States, 10–15% of the total number of firms creates 50–60% of jobs. The same high results were noted in Canada, France, the Netherlands, Spain and other countries of the world. Despite the main problem of RGC – the occasional nature of periods of high growth rates – the high net contribution to job creation makes them the object of close attention of regulators in many countries of the world. RGC creates associated effects by influencing the demand for products from related industries or creating opportunities for increased production and efficiency in product chains. However, expanding business in the digital economy necessarily requires additional external financing, which, as a rule, is difficult to access, especially for young companies that are only trying to occupy their niche in the market. In this regard, the differentiated mechanism of state support measures should be implemented on the basis of high-quality identification of RGC.

Methodology

Currently, the problem of effective state support for business is of primary importance for modern society (Fomina, 2018). As world experience shows, improving the quality of life, growth in wealth directly depend on the development of the business sector, the quality of the transformations being carried out in the digital economy,

---

Elena A. Fomina, Ilya I. Beloliptsev, Denis V. Chuvilin, Ufa branch of the Financial University under the Government of the Russian Federation, Ufa, Russia
Julia V. Khodkovskaya, Ufa State Petroleum Technical University, Ufa, Russia

https://doi.org/10.1515/9783110654486-001

which requires the use of effective tools (Fomina, 2018) state support for business, including RGC. The presence of digital technologies helps companies make operational decisions in order to increase the utilization rate of assets, reduce current costs and increase overall efficiency, and, therefore, the conditions for the growth and development of companies (Khodkovskaya, 2018). As a rule, RGC support programs are considered as an integral part of state influence on small and medium-sized enterprises (SMEs). However, government support measures in most countries are provided to SMEs, regardless of their RGC affiliation.

The methodological basis for establishing RGC identification criteria was the studies of D. Birch, OECD, the World Bank, etc. After studying the best Russian and foreign practices to determine the essence of RGC and the criteria for their identification, three approaches were identified: "absolute" (OECD criteria), "relative" (Birch index, DHS criterion) and "distributive" (Halvarson criterion). A systematic analysis of modern scientific papers (more than 20 authors) (Henreksonand, 2010) revealed the main signs of DBK growth (Baranova, 2016): indicator, measures, regularity, threshold value of the measure, process and demographics of companies. To identify DBK growth indicators, the economic growth models of companies are systematized (Poh, 2013), the analysis of which allows us to conclude that there is no typical RGC image, and the term "rapid growth" itself is a multidimensional and multifaceted object for study; Methodological approaches should be applied in a comprehensive manner using a multi-criterion RGC selection mechanism. The process of identifying RGC according to different countries shows that there is no relationship between the level of development of the country (for example, in terms of per capita income) and the share of RGC in the total number of companies, which poses additional tasks for researchers to determine how different RGC identification criteria are related to other indicators of economic growth (except for company income and number of employees). Therefore, it is advisable to structure the RGC identification process to establish the need for state support measures by creating a single vertically integrated RGC selection system based on a scientifically sound choice of growth indicators.

Results

A critical review of RGC approaches revealed the advantages of the OECD criteria and the Birch index, which is evident in the ease of application and comparability of results when comparing RGC lists across regions and countries. It should be noted that many researchers focus on their search among young companies and startups. An analysis of the reasons for the greater concentration of RGC among them shows that the main ones are the opportunities to master new markets, technologies and products, as well as the specific nature of competition in new market segments. In all countries, the majority of RGC is concentrated in a group of companies no longer than 5 years old, and in a number of countries (Brazil, Hungary, Côte d'Ivoire) the share exceeds 60%. Thus, it can be concluded that in most countries about 40% of RGC is accounted for by startups 1–2 years old. Among companies over 21 years old, the share of RGC practically does not exceed 20% (with the exception of Ethiopia – 22%). Despite the fact that the growth potential is really most often higher among young companies, this does not always mean that RGC is concentrated among small and medium-sized firms. World Bank research shows that RGC can be found among all categories of company size. In general, a review of RGC identification results using OECD criteria shows that their share in the total number of companies in developed countries varies from less than 2% (Austria, Germany, Italy, the Netherlands), to 6% (Finland, Sweden, UK, USA, Spain), to 10% (South Korea). Studies also show that using the OECD criterion (income) and the requirement of a minimum staff of 15 people, on average, the share of RGC is about 5–6% of the total number of companies. Moreover, such a conclusion is the same for developed and developing countries (African countries) (Goswami, 2019). Comparison of RGC samples obtained using different criteria leads to the need to solve the problem of taking into account the "multifactorial" economic growth in identifying RGC. In such a way, in the work (Delmar, 2013), based on the analysis of 19 growth indicators over a 10-year period, it was shown that company growth can manifest itself in different ways, in different indicators. In this regard, the use of one criterion as a measure limits the selection of RGC to only one possible option. Therefore, we also consider it possible to use the Birch index as a criterion, significantly expanding the sample of RGC compared to the OECD criterion (number). Another "stratum" in the analysis of DBK identification criteria and factors is market and industry analysis. The characteristics of market formation can influence the growth of the amount of RGC, for example, through concentration/deconcentration processes. Therefore, the concentration of the market on the principle of "grow or leave" leads to the fact that enlarging companies displace their less effective competitors, occupying their market shares and increasing their own growth rates (Kay, 2019). Reverse processes can lead to an increase in the number of companies in high-yield market segments, which causes increased competition and distribution of the overall growth potential among many companies and, as a result, a decrease in the number of potential RGC. The industry analysis of RGC composition emphasizes that individual business areas may be at different stages of the life cycle (Volovikov, 2015) and, therefore, have different growth potentials. At the same time, the widespread opinion about the greater likelihood of RGC in high-tech industries is not unequivocally confirmed (Goswami, 2019). However, these two aspects of RGC identification analysis remain little understood to date. Thus, the definitions and criteria for identifying RGC can vary significantly, focusing on certain aspects reflecting the goals and objectives of a particular study (Audretsch, 2012). In researches of social and economic consequences of activity of RGC the important place is taken by the "effects of a modulation" (spillover effects) considering influence of RGC on indicators of growth of the relevant industry or the region. It is noted that in a competitive market, the redistribution of resources takes place in favor of RGC, which leads to the departure of inefficient companies from the market and an increase in the efficiency of the industry as a whole due to an increase in the scale of RGC activities. Based on the key features of RGC (steady growth and identification indicators), the RGC identification algorithm was formed (Figure 1.1), the RGC grouping was proposed: resource-oriented (according to the number growth indicator), market-oriented (according to the revenue growth indicator), focused on current results (according to the profitability growth indicator), oriented on development (according to the growth indicator of net assets).

The presented algorithm as a whole makes it possible to solve the problem of taking into account the sustainability of fast growth rates of companies, and, therefore, identify RGC. However, the results obtained at this stage of the present study are not fully adequate to the task of differentiating companies in order to implement government support measures. The development of the crisis of globalization is increasingly bringing state regulation of the national economy to mandatory harmonization of aspects of fiscal and social policies, and a return to public values as national values. The priority task of the state is to support companies, projects that provide increased social significance. In this aspect, the social return on business, its participation in social entrepreneurship projects, the creation of public goods, social value – factors of stabilization of the social environment, consciousness are of particular relevance. The specificity of the "good" component is determined through the creation of Average Wage, the provision of social and service services that have the properties of environmental friendliness, safety, general accessibility for consumers, etc., which determine public value and accelerate positive changes and transformations in public life. In our study, the conce.pt of social value (Meynhardt, 2008) is expressed through the evaluation components of "good" and "anti-blago," in which social specificity is manifested. The manifestation of the "antiblag" is accompanied by a negative impact on the environment, quality of life and social stability, and an increase in macroeconomic and social tensions, which are subjects of government regulation. This means that almost all aspects of government regulation should be focused on solving socio-economic problems, taking into account the criterion of "public value." The effects of the "public value" criterion vary in RGC categories depending on the industry, age, innovative activity and other characteristics, therefore, the algorithm compiled by the authors (Figure 1.1) is advisable to supplement the 3rd stage (Figure 1.2) for the formation of a RGC sample using the "public value" criterion, having formed a single vertically integrated RGC selection system for providing state support measures.

The main indicators in the algorithm (Figure 1.2) are the scope of SSP and social stability of the enterprise. The feasibility of introducing these indicators is due to the need to provide priority measures of targeted support from the budget of the RGC, which implements projects in the field of social entrepreneurship. Systematizing the key areas of implementation of RGC social entrepreneurship projects in Russia, the

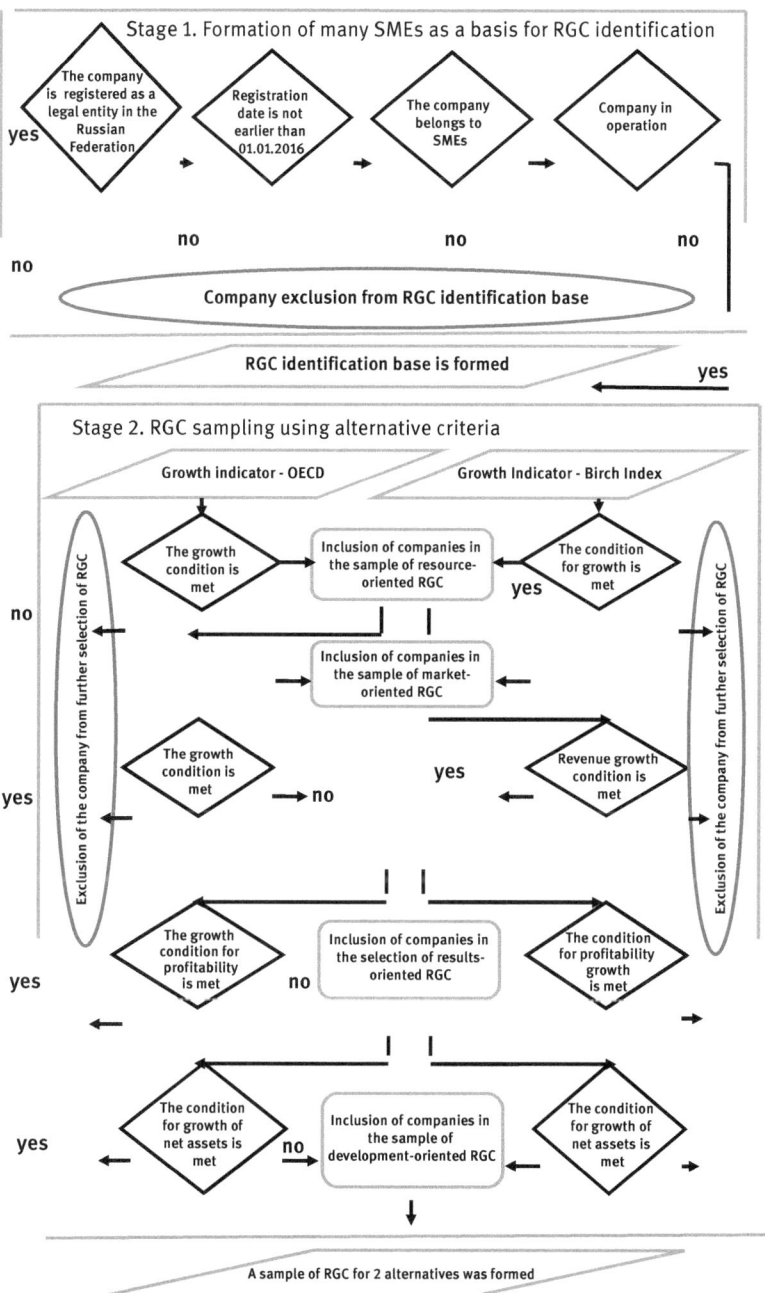

Figure 1.1: RGC identification algorithm.
Source: developed and compiled by the authors.

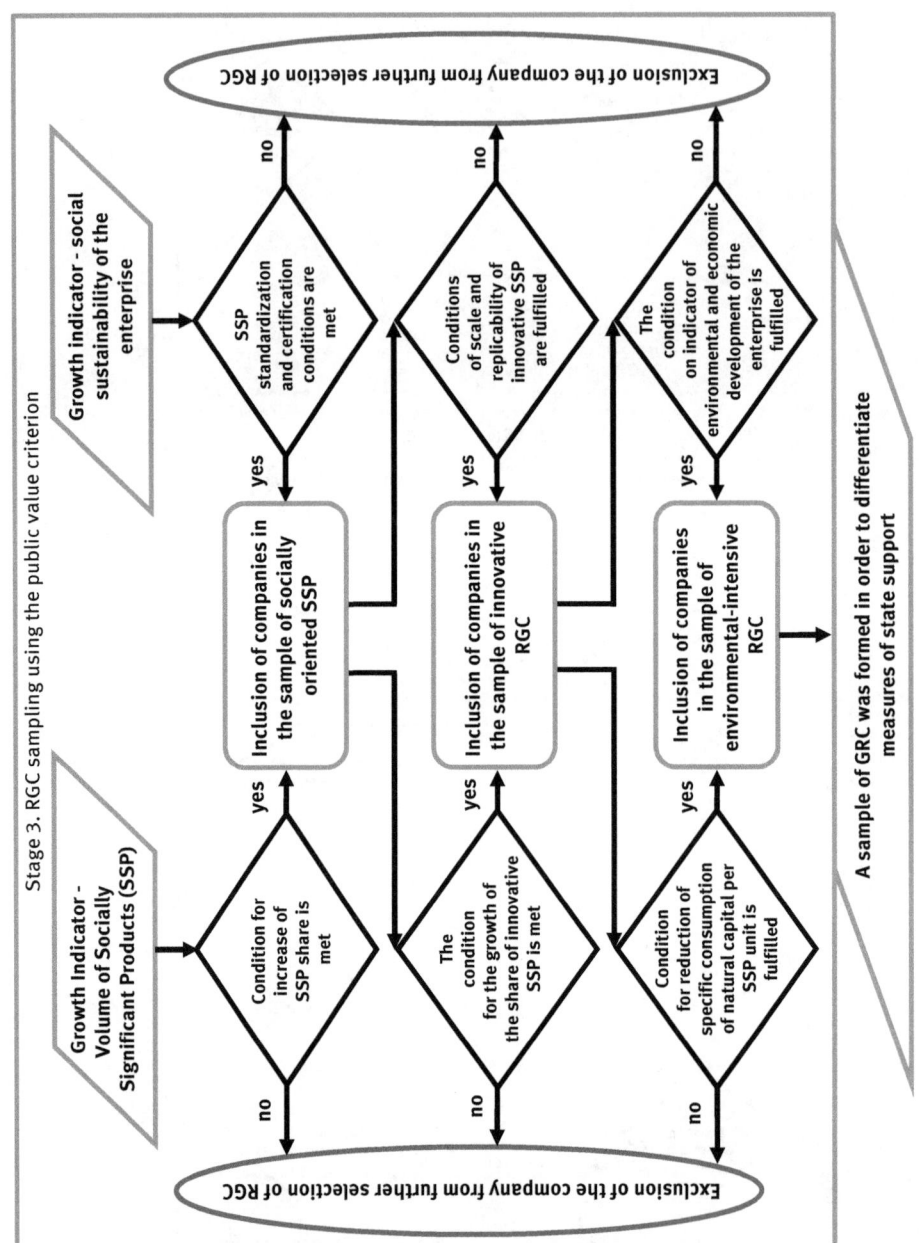

Figure 1.2: GRC selection algorithm for differentiation of state support measures.
Source: developed and compiled by the authors.

algorithm presents a grouping of three types of companies taking into account the criterion of "public value": socially oriented, innovative-oriented and nature-intensive.

The importance of providing state support measures to socially oriented RGC is explained by the fact that these companies are aimed at producing SSP that meets the requirements of international and national quality standards. Since the production of a certified RGC is a condition for improving the quality of life, ensuring national security, and preserving state sovereignty.

The transition to digitalization predetermined the priority of providing state support measures to innovative-oriented RGC, whose products are not only socially significant, but also feature innovative qualities. Innovative SSPs not only provide competitive advantages for RGC and contribute to increasing the pace of R&D deliverables, but also form a society of balanced needs as the goal of social policy of any state, based on the criterion of "public value." In order for government support measures to be effective, it is necessary to fulfill the following conditions for the production of innovative SSP – scalability and replicability, which determine the availability of goods and services for each consumer.

For national economies, solving environmental problems, efficient use of natural capital in conditions of limited resources is the most important task, therefore, natural-intensive RGC are identified as a separate type in the algorithm. the main reason for their allocation is the high environmental intensity of the Russian economy and the possible shortage of natural resources, noted by a number of experts. Thus, by stimulating the allocation of financial resources for the support and development of environmental-intensive RGC, the state slows down the depletion of the country's natural capital, stimulates the transition to intensive economic growth.

Despite the promotion of social values, national economies do not have a reference point for the implementation of the criterion of "social value." We believe that the differentiation of measures of state support for RGC taking into account the criterion of "public value" should be based on the recommendations of priority national projects, creating additional incentives for the development of companies in the digital economy, strengthening the state's activity in promoting the social component of RGC.

Conclusion

The presented study made it possible to draw a number of conclusions: the criteria for assigning companies to RGC may differ significantly depending on the goals of their identification and selection, but the main ones are the OECD criteria and the Birch index. As direct socio-economic effects, except for employment, it is customary to assess the contribution of RGC to the growth of total income of all operating companies. RGC is characterized by "overflow effects" or associated effects, which are observed both in the RGC presence industries (horizontal) and in other industries included in the common value chain (vertical). The problem of assessing the socio-economic results of RGC is the volatility of growth, therefore, the emphasis in identifying RGC is mixed with retrospective analysis to prospective growth forecasting. To differentiate measures of state support for RGC, it is advisable to use the criterion of "public value," which shows the involvement and return of business in the implementation of social projects focused on meeting public needs. A grouping of RGC was proposed, providing a synergistic effect of private-state partnership in the field of social entrepreneurship, a new quality of the created SSPs. In implementing state support measures for RGC, the state should provide priority assistance, create conditions for advanced economic development, the possibility of using modern achievements in science and technology to those companies whose contribution to the creation of socially significant values is accompanied by sustainable growth.

References

Audretsch, David B. (2012), "Determinants of High-Growth Entrepreneurship", Report prepared for the OECD/DBA International Workshop on – High-growth firms: local policies and local determinants, Copenhagen, 28 March.

Baranova, E.I. (2016), "The Study of Fast-Growing Companies – a New Tool for Microeconomic Analysis," World of the New Economy, No. 4, p. 98–104.

Daunfeldt, S.-O., Elert, N. and Johansson, D. (2016), "Are high-growth firms overrepresented in high-tech industries?", Industrial and Corporate Change, Vol. 25, No. 1, pp. 1–21.

Decker, R., Haltiwanger, J., Jarmin, R. and Miranda, J. (2014), "The Role of Entrepreneurship in US Job Creation and Economic Dynamism", Journal of Economic Perspectives, Vol. 3, No. 28, pp. 3–24.

Delmar, F., Davidsson, P. and Gartner, W. (2003), "Arriving at the highgrowth firm", Journal of Business Venturing, Vol. 2, No. 18, pp. 189–216.

Delmar, F., McKelvie, A. and Wennberg, K. (2013), "Untangling the relationships among growth, profitability and survival in new firms", Technovation, Vol. 8, No. 33, pp. 276–291.

Fomina, E.A., Khodkovskaya, Yu.V. and Kislitsyna, O.A. (2018), "Efficiency of state support for the development of small businesses," Eurasian Law Journal, No. 11 (126), p. 428–430.

Goswami, A.G., Medvedev, D. and Olafsen, E. (2019), "High-Growth Firms: Facts, Fiction, and Policy Options for Emerging Economies", International Bank for Reconstruction and Development, The World Bank, p. 193.

Haltiwanger, J., Jarmin, R.S., Kulick, R.B. and Miranda, J. (2016), "HighGrowth Firms: Contribution to Job, Output and Productivity Growth", US Census Bureau Center for Economic Studies Paper No. CES-WP-16-49, p. 75, 15 November.

Halvarsson, D. (2013), "Identifying High-Growth Firms", Ratio Working Paper, No. 215, Stockholm, Sweden, p. 32.

Henrekson, M. and Johansson, D. (2010), "Gazelles as job creators: a survey and interpretation of the evidence", Small Bus Econ, Springer, Vol. 2, No. 35, pp. 227–244.

Kaya, M.C. and Persson, L. (2019), "A theory of gazelle growth: Competition, venture capital finance and policy", available at: https://ideas.repec.org/p/hhs/iuiwop/1291.html (accessed 14 May 2020).

Khodkovskaya, Yu.V., Burlyakov, M.P. and Karabolatov, B.K. (2018), "Digital technologies as a factor in the growth of competitiveness and efficiency of the oil and gas business," Eurasian Legal Journal, No. 12 (127), p. 474–476.

Meynhardt, T. (2008), "Public Value: Oder was heißt Wertschöpfung zum Gemeinwohl?",der moderne staat – Zeitschrift für Public Policy. Recht und Management, No. 1, pp. 457–468.

Poh, Y. Ng (2013), "Characterizing high-growth firms: perspectives from the Asia-pacific region", A thesis submitted in fulfilment of the requirements for the Degree of Doctor of Philosophy in Management. University of Canterbury, p. 304.

Volovikov, B.P. (2015), "Assessment of strategic sustainability of the portfolio of projects based on dynamic models," Strategic Management, No. 1, p. 58–68.

Konstantin N. Lebedev, Yuliya I. Budovich
and Imomnazar E. Tursunov
2 Economy without Finance: Situation in R&D

Introduction

All financial systems for providing money (tax, credit, insurance, etc.) have substitutes from the real sphere of the economy, and the latter, obviously, as more acceptable for the kinds of activity they serve, displace the first (Lebedev, 2018a, Budovich, 2019a). Thus, the "normal" consumer credit is replaced by the "Islamic" (delivery of goods to the consumer by a Bank with deferred payment), the credit for the purchase of equipment – by equipment leasing offered by banking structures. This trend corresponds to the idea of secondary nature of financial measures to influence the economy, their effectiveness only in the case of taking in line with appropriate non-financial measures to solve social and economic problems (Lebedev, 2018-2, 2019-1, 2019-2, Budovich, 2019-2). The result of this trend is the formation of an economy in which the place of Finance is reduced to the necessary minimum. This economy, as well as the corresponding studies, we called "Economy without Finance", or "non-Financial economy". The purpose of the article is to show that there is a tendency to replace Finance with non-financial resources in the field of research and experimental development (R&D).

Methodology

Finance is understood as monetary relations that are not related to the oncoming movement of goods (products of labor created for sale) and other real assets or rights to use them.

Among the systems of supply there are both systems for providing funders (customers of products) of kinds of activity (systems for providing the funding), for example taxes as the system for providing the funding of prevention and liquidation of emergency situations, and systems for providing performers, including subcontractors (systems for providing the performance), for example the payment of financial assistance to the people affected by the flood, as a system for ensuring proper emergency response activities carried out by households (the former may

---

Konstantin N. Lebedev, Yuliya I. Budovich, Financial University under the Government of the Russian Federation, Moscow, Russia
Imomnazar E. Tursunov, Karshi engineering economics institute, Karshi, Uzbekistan

https://doi.org/10.1515/9783110654486-002

coincide with the latter, for example, when a household liquidates an emergency at its own expense).

Both types of systems can be financial and non-financial, for example non-financial alternative to the above payment was the free distribution of property of the first necessity (food, water, warm clothes, medicines), and of the reserve Fund of the Government of the Russian Federation as a system to provide funding for the elimination of the emergency situation – collection of the above tools and necessities declared by the governor of the affected region.

Initial security systems are also identified, such as revenue from the sale of goods (in the part of depreciation) as a source of funds for enterprises to replace worn-out equipment, and intermediate systems, such as long-term financial investments, into which the corresponding part of the revenue can be transformed before the date of replacement of equipment.

The article considers the relationship between the initial financial and non-financial systems for providing the funding and the performance of R&D.

The R&D statistics do not contain direct data on various financial and non-financial R&D supply systems. Statistical agencies do not require organizations that fund and perform R&D to provide the necessary information for their development. So in the Russian form of state statistical observation № 2-science, filled in by performers of R&D (Ministry of economic development of Russia, 2018), received funding of intramural R&D is characterized only from the point of view of its original sources, for example own funds, Federal budget allocations, funds of business organizations, while the estimation of the place of the various systems of cash supply for providing the funding and performance of R&D also needs data on the forms of target funding of R&D (purchase, contribution to the company's authorized capital, contribution to joint activities, credit, etc.) and the origin of the funds of the sponsors of R&D expenditure (taxes, revenue from the sale of goods, non-purpose credit, etc.). In form № 2-science, information is not collected that allows us to judge the place of non-monetary non-financial supply systems in the provision of R&D performers, for example, the cost of using mega-science installations.

Therefore, the study was based on indirect statistical data and data from individual studies.

Results

The rationality of the structure for providing the funding of R&D is primarily discussed in terms of the ratio in gross domestic expenditure on R&D (GERD) between government funds and business enterprise sector funds that form the main flow of R&D funding (they are supplemented by foreign and other national sources). However, the "competition" between these two mixed financial and non-financial systems is a veiled form of competition for a share in the funding of R&D between the two main financial and non-financial systems for providing R&D funding. This is a financial tax system that forms the main part of the government funds that go to funding of intramural R&D expenditure, as can be seen from the share of taxes in state budget revenues. According to the paper (Enina, 2009), in the US, taxes make up 90% of budget revenues, in Germany-80, and in Japan-75, or on average 82% (90 + 80 + 75) / 3. This is a non-financial system for providing the funding of R&D, such as the revenue of production companies from the sale of goods (in parts of depreciation charges and profits left at the disposal of the enterprise), which forms the main part of business funds that go to funding of intramural R&D expenditure, as can be judged by the share of depreciation charges and profits in company's own funds allocated for financing innovative development. According to the work (Mukhamedyarov, 2008), depreciation and profit remaining at the disposal of the enterprise, in the own funds of enterprises directed to financing innovative development, are more than 50 and more than 30%, respectively, i.e. more than 80% (50 + 30) together.

As it turns out, in the countries that are leading in scientific and technological development, half a century ago, the process of replacing the then dominant tax system with the revenue of production companies from the sale of goods began. This is evidenced by the loss of government funds of the dominant position in providing the funding of GERD. Thus, in the United States, public funding of GERD decreased from 57% in 1970 to 35 in 1994 and 27.1 in 2009, in the United Kingdom- from 54% in 1975 to 34 in 1995 and 30.7 in 2009 (Sudarikov, Gribovsky, 2012). In 2017 the ratio between government and business funds in GERD was 25.1% to 62.3% in the United States, 28.5% to 65.2 in Germany, 15.0% to 78.1 in Japan, 20.0% to 76.1 in China, and 22.7% to 75.4 in Korea. In Russia, this ratio was 66.2% to 30.2% (Gorodnikova et al., 2019), i.e. it was the opposite. Based on the fact that taxes form the main share of state funds, and the revenue of production companies – the main share of business funds that go to funding of intramural R&D expenditure, according to the data provided, we can conclude that in advanced countries, the revenue of production companies in GERD takes a significantly larger share than taxes.

Despite the fact that countries that account for the majority of the world's R&D expenditure and of its funding have achieved significant levels of substitution of government funds in GERD with business funds (see above), the global process of replacing taxes in funding of GERD by the revenue of production companies has significant potential, as shown by the following:

1. The process of large-scale replacement of government funds with the funds of the business enterprise sector in the funding of GERD has not been joined or has not been fully joined by some promising countries in terms of innovative development, and those that set a corresponding goal. This is primarily Russia (Kladova A. et al., 2018), as is clear from the above data. And about the relevant purpose the strategies of scientific and technological development of the Russian

Federation say, the first of which assumed in 2015 to achieve 70% share in the funding of GERD of extrabudgetary sources (Interdepartmental Commission on science and innovation policy, 2006), i.e. reducing the share of budgetary sources to 30% (100–70).
2. Many countries that are leaders in scientific and technological development have not yet reached their national or, more significantly, their recommended goals for GERD as a percentage of GDP, for example, European countries and the United States – a 3 percent share that was recognized as desirable for Europe in 2002, and for the United States in 2009 (Lanshina, 2017). In Germany, GERD in GDP in 2017 was 2.93%, in France – 2.25%, in the United Kingdom – 1.69%, and in the United States 2.74 (Gorodnikova et al., 2019), i.e., among the largest Western economies, only Germany has almost reached the target level of 3%.

In addition, there is a high probability that the target levels will be revised upwards.

In this case, the opinion is presented, for example, in the work (David et al., 2000) that the increase in the level of government funding for R&D is inappropriate (it displaces private funding, leads to higher R&D costs primarily due to the increase in wage of researchers that encourages the private sector to reduce the relevant expenditure, etc.). Therefore, the expected increase of GERD in GDP target levels is likely to occur due to an increase in the share of business funds in GERD.

Thus, in the world the process of replacing the tax system with such a non-financial substitute as the revenue of production companies from the sale of goods has been going on for half a century and, obviously, is for a long time to come.

The replacement of financial systems by non-financial ones in R&D also occurs in the part of systems for providing R&D performance. Frascati Manual 2015 (standards in R&D statistics) divides all external funding received by R&D performers into transfer funds (grants, crowd-funding, debt forgiveness, philanthropy, etc.) and exchange funds, i.e. financing R&D in return for its performance (R&D purchases, R&D outsourcing, contributions to collaborative R&D, etc.) (OECD, 2015). As noted in the OECD document (Aspden, 2008), data received from countries around the world that divide external R&D funds into grants and acquisitions suggest that the volume of acquisitions is, first, quite significant and, second, growing.

Conclusion

The process of replacing Finance with non-Finance is also observed in R&D. Relevant researches are seriously hampered by the lack of statistical data. In the face of serious challenges from the financial sector, statistical agencies should provide governments, businesses, analysts and the public with information about the place of various financial and non-financial supply systems in economic activity.

References

Aspden, Chr. (2008). Issues of updating the 1993 SNA. Research and development. OECD, Geneva, 24 p., available at: www.unece.org (accessed 19 December 2019).

Budovich, J.I. (2019a). Crowdfunding in the mirror of the non-financial economy. Business. Education. Law. Bulletin of Volgograd Business Institute, February, 1 (46), pp. 40–46.

Budovich, J.I. (2019b). Promising non-financial tools for solving socio-economic problems. VESTNIK TVGU, 4 (48), pp. 243–246.

David P.A., Hall B.H., Toole A.A. (2000). Is public R&D a complement or substitute for private R&D? A review of the econometric evidence. Research Policy, 29 (4–5), pp. 497–529.

Enina, P.E. (2009). Taxation. GOUVPO Voronezh State Technical University, Voronezh, available at https://historich.ru/uchebnoe-posobie-voronej-2009-gouvpo-voronejskij-gosudarstvenn/index2.html (accessed 10 December 2019).

Gorodnikova N.V., Gokhberg L.M., Ditkovsky K.A. and others. (2019). Science. Technologies. Innovation: 2019: brief statistical compendium. SRU HSE, Moscow. 84 p.

Interdepartmental Commission on science and innovation policy. (2006). Strategy for the development of science and innovation in the Russian Federation for the period up to 2015. Protocol No. 1 of February 15. Legal Reference System ConsultantPlus, available at: http://www.consultant.ru (accessed 08.01.2020).

Kladova A., Alpidovskaya M., Gordeev V. (2018) The Shift of the Competition Paradigm in the Banking Sector of Russia. The Future of the Global Financial System: Downfall or Harmony. Book series: Lecture Notes in Networks and Systems, vol. 57. Springer, Cham, pp. 61–68.

Lanshina, T.A. (2017). Innovative Sector of the USA: State Policy and Tendencies of the Last Years. Management consulting, 6, pp. 73–87.

Lebedev, K.N. (2019). About the effectiveness of financial and non-financial measures for solving problems of R&D in Russia. Innovative economy: prospects for development and improvement, 1 (35), pp. 185–196.

Lebedev, K.N. (2018a). Economy without Finance. Economic and Law Issues, 6 (120), pp. 45–53.

Lebedev, K.N. (2018b). "Key" role of finance in socio-economic development. Economic Sciences, 6 (163), pp. 7–15.

Lebedev, K.N. (2019). The agreement of the government with business on voluntary financing of social projects by excess profits as a new and promising institution of non-financial economy. Business. Education. Law. Bulletin of Volgograd Business Institute, February, 1 (46), pp. 69–76.

Ministry of economic development of the Russian Federation. Federal state statistics service. (2018). On approval of statistical tools for the organization of Federal statistical monitoring of activities in the field of education, science, innovation and information technology. Order No. 487 of August 6. Legal Reference System ConsultantPlus, available at: http://www.consultant.ru (accessed 03.06.2019).

Mukhamedyarov, A.M. (2008). Innovative management. INFRA-M, Moscow, 137 p.

OECD (2015). Frascati Manual 2015: Guidelines for Collecting and Reporting Data on Research and Experimental Development, The Measurement of Scientific, Technological and Innovation Activities. OECD publishing, Paris, available at DOI: http://dx.doi.org/10.1787/9789264239012-en (accessed 27.11.2019).

Sudarikov, A.L., Gribovsky, A.B. (2012). Public-private partnerships in science, technology and innovation: analysis of the international experience. Innovations, 7 (165), pp. 47–59.

Gulnora Sh. Karabaeva, Rano R. Nazarova
and Gulchexra N. Nigmatullayeva
3 Innovative Activity in Post-Pandemic Conditions

Introduction

The modern stage of social and economic formation of Uzbekistan is characterized by increased competition between manufacturers of the real sector of the economy, one of the main tools of which are the price and quality characteristics of goods and services. The presence of a competitive superiority of the enterprise in the dominant order is determined by the limited range of offered products and services. For this reason, the creation of a mechanism for improving the organization of the production process, which allows the company to use all possible alternatives to achieve the main goal of the activity – maximizing profits – is of particular importance.

In the current situation in the world, there is a transformation of economic models of countries, the cornerstone of which has become the activation of the process of innovation. This phenomenon involves the application of foreign experience, research, manufacture and sale of goods, services and technology-intensive production, as well as the improvement and implementation of modern production technologies.

In a changing environment, increasing competition between producers, the primary task of industrial enterprises is the organization of the release of innovative products and services. The procedure for introducing innovations is characterized by the volumetric structure of the elements involved, with the mandatory inclusion of the results of scientific and technological progress.

Materials and Methods

The study used methods of analysis and synthesis, induction and deduction, system analysis, comparative analysis, indicative selective supervision, comparison, correlation and regression analysis, economic and mathematical modeling.

The problem of innovative development of the economy has become urgent in recent years and occupies a leading position in the works of foreign and domestic scientists. One of the first issues of the innovative economy was dealt with by

---

Gulnora Sh. Karabaeva, Tashkent Branch of Russian Economy University named after G.V. Plekhanov, Tashkent, Uzbekistan
Rano R. Nazarova, Gulchexra N. Nigmatullayeva, Tashkent state university of economics, Tashkent, Uzbekistan

https://doi.org/10.1515/9783110654486-003

Schumpeter (1982), who actually was the founder of the theory of innovative development. P.A. deals with innovation management issues Fathutdinov (2002), Morozov (1998), Baryutin (2004).

Issues related to the analysis of innovative processes were investigated in the works of such authors as Abramov (2000), Wootton (2014), Yakushev and Dubynina (2017) and others. At the same time, the issues of innovation and the development of industrial enterprises in the aggregate remain insufficiently studied in the scientific literature.

Results

Coronavirus does not sort people by their nationality, religion or citizenship. Negative influence is felt everywhere and everywhere. Although healthcare workers and technicians are in first place today, strategic development is still linked to the activities of economists who forecast rising unemployment and declining economic indicators in many countries. These expected problems are pushing society and the state to rebuild in order to get the maximum benefit under the given circumstances. In modern conditions, the formation of a new model of economic development of states, the most important characteristic of which was the activation of the process of innovation. This direction includes both the use of the advantages of the international division of labor, the development, production and export of domestic high-tech goods, services and technologies, as well as the development of new advanced production technologies available abroad.

The study of foreign experience in the formation of national innovation systems shows that without the participation of the state it is impossible to implement innovative programs in economic sectors. In the process of industrial development of society, the state took upon itself the training of personnel and the creation of appropriate infrastructure. In the transition to the innovation phase, it takes on the organization and financing of the economic sector, which serves as the basis for the interaction of science, education and business. Innovation is a product of such interaction and a new economic resource actively involved in economic development.

A comprehensive assessment of innovative potential should cover the resources involved at all stages of the innovation process, and should not be limited to indicators of the stage of production and distribution of innovations.

The diagnostic approach to assessing innovative potential covers the enterprise as a whole. The evaluation of the internal component of the system is determined relative to similar systems, i.e. competing enterprises. Therefore, in this case, the scope of strategic analysis is the analysis of the competitive position of the enterprise. For this, both analytical models (SWOT analysis) and competitive analysis models (strategic management and marketing) can be used.

Personnel indicators, in our opinion, should characterize the security of the innovation process with human resources, the qualification and age structure of the personnel involved in the creation and dissemination of innovations. Analysis of the information provided by the statistics showed that the assessment of personnel involved in innovative processes is carried out mainly by quantitative indicators.

Currently, the problem of "aging scientific personnel" is quite relevant for industrial enterprises. The average age of staff tends to decrease steadily. We consider it necessary to include an indicator of the age of the personnel involved in innovation processes in the system of indicators of innovative potential. The validity of this decision is as follows: for young specialists to master the necessary skills and study the specifics of production processes in order to acquire high qualifications and skills, it takes time, sometimes calculated over the years. The company's management should pay special attention to this problem and the most modern methods of labor motivation should be used to attract young specialists.

Assessment of innovative potential according to the proposed indicators characterizing the main resources of enterprises used in innovative activities will not be objective without assessing the effectiveness of their use. In other words, the availability of resources of enterprises of even the highest quality and in the required quantity does not mean that the enterprises make full use of the existing innovative potential. The approach to a comprehensive assessment of innovative potential, in our opinion, should be based on the fact that the goal of creating and accumulating the necessary resources for enterprises implementing innovations is, firstly, the ability to create innovations on a regular basis, and secondly, profit from implementation of innovative products. Thus, the system of indicators evaluating the innovative potential of enterprises should include indicators of the effectiveness and efficiency of innovation.

Analysis of production factors is carried out with the aim of forming innovative activity in increasing the competitiveness of the industrial sector, affecting the stages of strategic development.

The need for assessing the innovative activity of industry is manifested in the initial stage of implementing an innovative strategy, coupled with a specific innovation. Thus, innovative activity depends on industrial potential, is compared with a particular innovative project in order to reveal the ability of an industrial enterprise to implement innovations. With a full analysis of innovative potential, the method of checking the internal environment of the enterprise is applied.

The degree of elements of innovative potential in the proposed methodology is determined in a single way, which is based on the established overall indicator.

A general analysis of innovative opportunities necessarily includes the means used absolutely in all stages of the innovation cycle, is not limited to time, stage and implementation of novelty.

Diagnostics of innovative potential is carried out in all industrial areas. The content of the term "innovative potential" is determined by the classification of indicators reflecting the modern formation of the production concept. In the enterprise development strategy, a comparative characterization of innovative processes in the regional and foreign markets should be carried out.

In the process of our study, we found that the following indicators are used to assess the innovative potential of the enterprise:
- staff innovation (number, quality of specialists and structural organization of the enterprise)
- material and technical support (expenses for scientific research, scientific research and development work, commercialization of ideas; technological level of the production base)
- information element (availability of information databases, quality of information distribution, user satisfaction)
- summary indicators of innovation (Medynsky, 2018)

In order to determine the production efficiency of one of the country's leading industries – industry, the impact of expendable resources on gross output was studied in 2000–2019. Information is given in Table 3.1.

Table 3.1: Dynamics of the main indicators of industry of the Republic of Uzbekistan for 2000–2019.

Years	Gross Industrial Product Cost, billion soums, Y	Cost of fixed assets, billion soums, K	Number of production and industrial personnel, thousand person, L
2000	1,888.9	638.8	1,145.0
2001	2,830.8	3,085.1	1,160.0
2002	4,494.0	5,371.6	1,186.0
2003	6,127.5	6,469.8	1,223.3
2004	8,123.2	7,720.9	1,283.9
2005	11,028.6	9,133.5	1,347.5
2006	14,640.3	11,094.3	1,402.4
2007	18,447.6	13,753.6	1,445.5
2008	23,848.0	16,638.6	1,486.7
2009	28,387.3	21,130.9	1,513.1
2010	38,119.0	25,454.1	1,539.6
2011	47,587.1	31,090.7	1,563.7
2012	57,552.5	38,718.1	1,589.9

Table 3.1 (continued)

Years	Gross Industrial Product Cost, billion soums, Y	Cost of fixed assets, billion soums, K	Number of production and industrial personnel, thousand person, L
2013	70,634.8	48,608.9	1,615.3
2014	84,011.6	58,961.6	1,642.3
2015	97,598.2	68,185.6	1,668.2
2016	111,869.4	94,898.7	1,802.4
2017	148,816.0	115,659.6	1,826.8
2018	235,340.7	126,547.7	1,802.9
2019	331,006.6	200,787.4	1,820.5

Source: developed and compiled by the authors by (Medynsky, 2018).

Based on the data in Table 3.1, we construct the production function of the industry of the Republic of Uzbekistan:

$$Y = 0.029 * K^{0.3938} * L^{5.712} \tag{1}$$
$$(5.39) \ (0.09) \ (0.86)$$

Judging by the revealed dependence of the factors, it was determined that factors that we did not take into account influence the production function with a coefficient of 0.029.

The coefficient calculated on fixed assets of the industry equal to 0.3938 is the coefficient of elasticity and shows 0.3938 percent change in the gross industrial output caused by a 1 percent change in the value of fixed assets.

This means that a 1 percent change in the number of people employed in industry shows an average of 5.712 percent increase in gross output.

This means that the development of industry in our republic by 93.55 percent (an increase in those employed in industry) depends on extensive factors. And this requires a transition to the path of intensive development. This means that our studies of the state of the industrial sector and the parameters of the compiled production function of Cobb-Douglas show that it is necessary to diversify industrial production, develop on the basis of innovative technologies, and also develop a new concept for the model of effective functioning and innovative development of the republic's industry.

The cost of gross industrial products of the Republic of Uzbekistan from 2000 to 2017 was constantly growing. In 2017, compared with 2000, it increased by 140692.8 billion soums. or 18.3 times. Since 2018, growth has been accelerating.

This was due to the attraction of a large volume of investments in the industry based on programs for the prospective development of the industrial sector, marked by the objectives of the Strategy of Action for the five priority areas of the country's development in 2017–2021, the commissioning of new production facilities, etc.

Gross output of the Republic of Uzbekistan in the forecast period has a tendency to increase. The average increase in gross output is 20,278.3 trillion. sum This growth is 147.9 times more than in 2000 and 7.33 times more than in 2010.

This was mainly influenced by the introduction of innovative technologies in the country's industrial sectors, the use of new methods of organizing industrial production (cluster) and a number of other factors. The share of industrial production in gross domestic product is expected to be 42.06% in the forecast period (2023).

So, we conducted an analysis of factors affecting innovation activity in industry, and determined the following dynamics of industrial personnel.

The number of industrial personnel in 2000–2015 increased by an average of 35–40 thousand people. And as a result, in 2018 amounted to 1802.9 thousand people. In the forecast period, there is an annual increase in the number of industrial personnel by 52–55 thousand people. According to estimates, by 2023 the number of people employed in industry will reach 2128.0 thousand person (Figure 3.1).

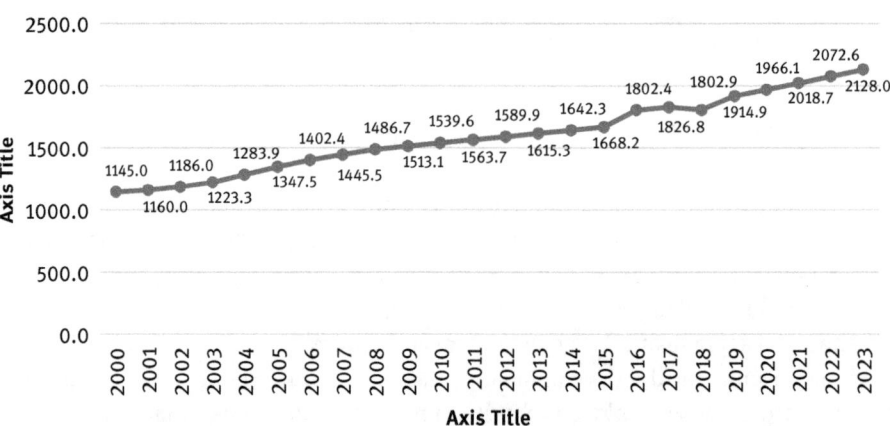

Figure 3.1: Dynamics of the number of people employed in the industry of the Republic of Uzbekistan for 2000–2019 and forecast values for 2020–2023, (thousand person).
Source: developed and compiled by the authors.

The main signs that reflect the internal environment are the quantitative and qualitative level of personnel. The degree of quality of personnel and other properties are interrelated with the scale and pace of the implementation of innovative activities.

We think that professional characteristics show the level of security of the innovation process with human resources, the qualification and age structure of personnel connected to the process of creating and disseminating innovations. According to the results of the analysis, revealed the frequent use of quantitative indicators of innovative personnel.

Nowadays, the problem of "aging scientific personnel" is manifested in industrial enterprises. The number of workers of moderate age is constantly decreasing. We believe that it is necessary to check the age of the staff working in the innovation unit. As a rule, new employees need training with additional knowledge and skills in order to acquire higher qualifications. This usually takes years. At enterprises, it is necessary to work with young personnel and at the same time develop new methods of labor motivation.

Of course, conducting research and development work in industry will lead to the introduction of innovations in the industry and an increase in the volume of gross output. The following Table 3.2 presents the indicators of innovation in the industrial sector of Uzbekistan for the analyzed period.

Table 3.2: Industrial Innovation Indicators 2009–2018.

Years	Industrial innovations (marketing and technological), pcs.	Number of R&D personnel, people	Costs of marketing and technological innovations, mln UZS	Industrial patents and inventions, pcs.
2009	231	220	276,313.8	37
2010	243	171	264,228.4	37
2011	301	108	367,401.0	35
2012	274	114	271,652.4	21
2013	684	165	3,123,516.8	25
2014	872	104	3,258,026.8	18
2015	855	98	5,524,566.6	11
2016	959	103	1,210,045.5	8
2017	1177	133	1,919,747.1	6
2018	1379	145	2,900,828.4	2

Source: developed and compiled by the authors.

It can be concluded that the number of marketing and technological innovations achieved relatively stable growth. In 2018, 5.9 times more innovations were introduced than in 2009. As for the cost of them, here we see an increase over this period of 10.5 times. Unfortunately, since 2014, the number of patents and inventions obtained has begun to decrease, which once again confirms the relevance of the development of innovative activity in the industrial sector of the economy.

Figure 3.1 shows the dynamics of the research and development work carried out in the industry for 2000–2019 and the forecast values for 2019–2023.

The value of the cost of research and development in industry in the forecast period annually is on average more than 1.29 trillion. soums. In 2023, the volume of expenses for research and development in relation to 2010 will increase by 15.6 times and amount to 1,901,533.1 million soums.

Until now, the number of patents and licenses has been indicators of the effectiveness of research and development within individual enterprises. In modern conditions, given the current downward trend for a number of reasons, the propensity to patent R&D results by most enterprises, the use of this indicator in assessing innovative potential is doubtful in accordance with realities.

It is known that the importance of each enterprise in the national economy is determined not only by its gross output, net profit or income, but also by the creation of new jobs. Solving the problem of employment, industrial enterprises contribute to the expansion of enterprise and initiative of personnel.

It is clear that industrial enterprises cannot work separately from each other, during the production process they cooperate with suppliers of raw materials, materials and other means of production, and communicate with consumers of products and services. And here, any failure associated with the continuous flow of raw materials and financial resources or equipment and other violations of business rules and agreements leads to a decline in the economic potential of the enterprise, to its insolvency. Consequently, we will not see the next stage of industrial progress until all the manufacturers have learned how to organize production and work in strict compliance with their obligations to both partners and the state.

Thus, for the innovative development of industry, it is necessary to conduct research in high-tech areas of industrial sectors. This, in turn, will ultimately lead to an increase in gross industrial output.

Assessment of innovative potential according to the proposed indicators characterizing the main resources of enterprises used in innovative activities will not be objective without assessing the effectiveness of their use. In other words, the availability of resources of enterprises of even the highest quality and in the required quantity does not mean that the enterprises make full use of the existing innovative potential. The approach to a comprehensive assessment of innovative potential, in our opinion, should be based on the fact that the goal of creating and accumulating the necessary resources for enterprises implementing innovations is, firstly, the ability to create innovations on a regular basis, and secondly, profit from implementation of innovative products. Thus, the system of indicators evaluating the innovative potential of enterprises should include indicators of the effectiveness and efficiency of innovation.

Uzbekistan has sufficient resources to pursue an active innovation policy – a luxurious mineral resource base, a diversified industrial complex and scientific and technical structures, a large contingent of scientific and engineering personnel and experienced practitioners. A number of industries have introduced unique advanced technologies and the possibility of establishing the production of science-intensive types of products. The Republic of Uzbekistan has a significant comparative advantage, which consists in the fact that the development of industries that manufacture products with high added value occurs not only due to the general state of technology or industry, but also due to the availability of actual high-quality human capital. The modern industrial staff and the scientific sector, with the directed support of the state, are capable of producing products that can not only satisfy the demand of the population, but can also be exported abroad and increase the country's economic level.

The Action Strategy for the five priority areas of development of the Republic of Uzbekistan in 2017–2021 defines specific tasks for the development of the innovative potential of industrial enterprises. Powerful industrial enterprises have been created and operate in Uzbekistan, representing almost all sectors of the economy – from heavy to light industry, from processing agricultural products to high-tech industries. In their activities they use all known factors of production – land, natural and labor resources, machinery and technology, investments, information technologies, which are the production potential or national wealth of the country.

The structural reforms carried out in the Republic of Uzbekistan are aimed at the speedy development of the industrial sector of the economy. However, calculations and studies show that the republic's industry is developing mainly in an extensive way. But, based on foreign experience, we found that the dynamic development of industry often occurs in an intensive way, i.e. development is driven by innovation.

In a situation of increasing globalization of the industrial producers' market, competition is becoming more intense. In order to calculate and strengthen the competitive properties of goods, when managing an enterprise, it is necessary to actively develop and introduce innovations, continuously improve existing ones and seek new development strategies, forms and methods of managing production activities aimed at quickly and fully satisfying the needs of the market.

In such conditions, an important factor for any enterprise in the market from the standpoint of increasing its competitive efficiency is the introduction and use of innovation in the creation and proposal of innovative products. Certainly, such a direction can develop an updated value for the products and services offered by customers, which can subsequently increase the competitiveness of the enterprise in the market.

Since the management of enterprises should plan their innovative activities, the question of the systematization of this work will certainly arise.

The current stage of the formation and development of innovative activity in Uzbekistan is associated with some problems. The republic has not yet succeeded in guaranteeing the effective use of a wide range of applied elements and means for the course of innovative processes, and the required level of complexity of their application is not achieved. There is a significant delay in the methodological support of the applied mechanisms for regulating the innovation process, as well as non-observance of intellectual property rights. It is also impossible to ensure comparability of key indicators of economic development (prices, interest credit rates, taxes, etc.) with general economic development trends, structure and technology levels of the real sector of the country's economy. Therefore, there is a relatively low share of innovative enterprises (about 1% of all enterprises), export of high-tech science products (about 11% of total imports), a significant lag behind developed countries in international ratings like the Doing Business, Economic, Freedom, Corruption Perception Index, lack of labor resources engaged in innovative research (about 1% of the employed population).

Discussion

The current position of the country's innovative development is determined by the complex interactions of scientific, technical and technological structures of the economy and the system of financial and economic relations. Significant factors in this are the deficit of equity capital of enterprises and attracted investments, insufficient concessional lending, fiscal focus on taxation, etc. They hinder the process of research and study of new types of technological equipment, materials and products used.

The implementation of the state policy for the development of the innovation sphere is permissible only if the role of the state in resolving issues of both a scientific, technical and organizational-production nature is strengthened. Transformation to systematic work on strengthening and developing scientific, technical and investment activities, updating the production structure, and enhancing entrepreneurial activity in the innovation sphere is possible at this stage only with the direct participation of the state.

The main trends in increasing the innovative activity and competitiveness of industrial enterprises are the phased and consistent implementation of organizational and production, financial, technological, interbranch, intraeconomic and foreign economic measures. To effectively use the existing potential, increase the competitiveness of domestic products, increase their exports and avoid unnecessary costs, systematic support is provided for representatives of the real sector of the economy. In the course of implementing the decisions of the head of our state, the number of modern enterprises and farms, specialized manufactures that use the most advanced equipment and technologies is growing, and the interaction of agricultural producers with enterprises in the processing sectors is being strengthened.

The intensification of innovation should be based on a set of scientifically based principles in the context of a rapid process of scientific and technological development and increasing consumer requirements for the quality characteristics of goods and services.

When intensifying the innovation activity of high-tech industries, an extensive stock of various methods should be used, which involve methods and techniques of enterprise management and impact on personnel in order to enhance innovation.

Over the years of independence, the structure of industrial production in Uzbekistan has changed in accordance with the requirements of a market economy, its progressiveness has intensified, which is reflected in the accelerated development of basic industries that determine scientific and technological progress and the fastest transfer of the economy to the tracks of modernization and renewal.

The main factor in accelerating structural transformations was the creation of a favorable investment climate in the republic. So, today 3 investment companies, 1 venture capital fund and 1 management company have been created. Active work is underway to attract venture capital to domestic startups, a program has been organized to train personnel and conduct further research in the field of venture financing in the country's economy.

In accordance with the Decree of the President of the Republic of Uzbekistan dated November 24, 2019 No. UP-5583 "On additional measures to improve the mechanisms for financing projects in the field of entrepreneurship and innovation", a List of priority innovative development projects and startup projects recommended by the Ministry of Innovative Development for venture financing, which provides for the implementation of projects in the amount of $ 28,205.3 thousand with a implementation period until 2026. Of these, 8 new innovative projects, 4 start-ups and 1 development project aimed at the development of industries, services and infrastructure.

The Concept of the Development Strategy of the Republic of Uzbekistan until 2035 provides for entry into the top 50 countries according to the Global Innovation index and to implement investment projects in the amount of 50.9–62.2 billion US dollars.

Conclusion

The prospects for the development of the scientific and technical complex of Uzbekistan will be largely determined by the commercialization of developments in the field of applied science, the creation of a full-fledged innovation market, and the ever-increasing orientation of the results of their achievements to the introduction into economic practice in the form of technology parks, business incubators and free economic zones.

The interests of our republic reaching new frontiers require the continuation of structural transformations and diversification of the economy. Only through the implementation of this cross-cutting task can we ensure the competitiveness of our country on the world stage.

References

Abramov, S.I. (2000). Investing. Moscow, TsEM.
Baryutin, LS (2004). Foundations of Innovation Management: Theory and Practice. Moscow, Economics.
Fatkhutdinov, RA (2002). Innovative management. St. Petersburg. Science.
Khakimov, Z.A. (2016). Factors that increase the competitiveness of light industry enterprises. Tashkent: Economics and Finance.
Medynsky, V.G. (2018). Innovation Management. Moscow, INFRA-M.
Morozov, V. (1998) Methods for assessing the quality of investment projects. The economist, 7(1), pp. 81–85.
Nikitin, V.V. (2016). Influence of innovative processes on competitiveness of regional economy (on the example of the Chuvash Republic). Humanitarian, social and economic and social sciences, 4(1), pp. 186–188.
Rasulev, A.F., Trostyanskiy, D.V., Islamova, O.A. (2015). The development of innovative potential and the trends of innovative activity of enterprises of Uzbekistan's industry. Economic newsletter Donbas, 2(40), pp. 49.
Schumpeter, J. (1982). The theory of economic development. Moscow, Progress.
Vinnikova I.S., Kuznetsova E.A., Repina R.V., Korovina E.A. (2016). Topical issues of innovative development enterprises of industrial sector of Russia. Internet-journal "NAUKOVENIE" 8(6), http://naukovedenie.ru/PDF/54EVN616.pdf.
Wootton, S. (2014). Strategic Planning: The Nine Step Programme. New York, Wootton,Home.
Yakushev, A.A., Dubinina, A.V. (2017). Innovation economy. Moscow, Finance and statistics.

Evgeniya K. Karpunina, Zulay K. Tavbulatova, Yuri V. Kuznetsov,
Naida D. Dzhabrailova and Olga A. Anichkina

4 The Challenges of Digitalization for Economic Relations of Tourism Industry Subjects

Introduction

The dynamics of the development of the tourism industry over a long period of time is characterized by stability and continuous growth (with the exception of periods of shocks and catastrophes). According to UNWTO, tourist flows in the world in 1950 were 25 million people, in 1980 they increased to 278 million people, in 2000 they intensified to 674 million people, and in 2019 they reached 1500 million people (UNWTO, 2017). The revenue generated by destination countries around the world from international tourism increased from 2 billion US dollars in 1950 to 1530 billion US dollars in 2019 (Tohology, 2020). According to the World Travel and Tourism Council (WTTC), the tourism industry in 2018 accounted for 10.4% of global GDP (8,811. 0 billion US dollars) and 319 million jobs, or 10% of total employment (WTTC, 2019). In the structure of total expenditures, 78.5% of the total volume is allocated to the leisure market, 21.5% – to business expenses.

If to track dynamics of development of the tourism industry on a global scale, we can see that the intensification of growth in the mid-20th century is associated with the emergence of mass institutional tourism, development of railway communication and transport infrastructure (Gyr, 2010). The rapid growth of the tourism industry at the end of the second millennium is due to informatization, increased openness and accessibility of information, and the strategic use of digital platforms as key competitive factors of the tourism industry (Murdaugh, 2005). In 2018, about 41% of business trips and 60% of tourist trips were planned and carried out using Internet services and applications (WTTC, 2019). However, the introduction of digital technologies is a challenge for the tourism industry: the ongoing changes in the needs of consumers of tourist products, new opportunities to meet them, as well as the use of digital solutions in all elements of the value chain are transforming the economic relations of the tourism industry (Poon, 1993). As a result, traditionally

---

Evgeniya K. Karpunina, Tambov State University named after G.R. Derzhavin, Tambov, Russia
Zulay K. Tavbulatova, Chechen state University, Grozny, Russia
Yuri V. Kuznetsov, Saint-Petersburg State University, Saint-Petersburg, Russia
Naida D. Dzhabrailova, Dagestan State University, Makhachkala, Russia
Olga A. Anichkina, K.G. Razumovsky Moscow State University of technologies and management (the First Cossack University), Moscow, Russia

effective travel companies, and often entire segments, lose their positions, they are forced to leave the market, incurring losses and leaving many employees unemployed.

Methodology

The theoretical basis of the research was scientific publications reflecting changes in the tourism industry under the influence of digital technologies. Thus, Poon (1993) emphasized that information and communication technologies spread throughout the tourism industry and affect all its subjects, since information is its driving force. The paradigm shift that changes the structure of the tourism industry is described in Xiang and Gretzel (2010), Cherevichko and Temyakova (2019), Bogomazova et al. (2019), Buhalis and Amaranggana (2013). Buhalis (1998) studied ways to meet the tourist demand and the survival of tourist companies in the long term. The advantages of digital technologies and services used in the tourism industry have been the subject of research by Coates (1993), Benckendorff et al. (2014), MacKay and Vogt (2012).

The purpose of the research: to reveal the specifics of the transformation of economic relations of tourism industry subjects under the influence of digital technologies, to identify the causes of changes and determine the prospects for their further development. Tasks: 1) Determining the structural features of the tourism industry as part of the economic system and identifying the nature of the impact of digital technologies on the economic relations of its subjects. 2) Systematization of digital technologies used in the tourism industry. 3) Justification of the reasons for changes in economic relations of tourism industry entities. 4) Determining the prospects for the development of economic relations of tourism industry entities in the conditions of digitalization.

To solve these tasks, the authors analyzed digital technologies used in the tourism industry based on a review of open sources and scientific literature, as well as the application of the method of systematization. Analysis and synthesis methods, comparison methods, logical and dialectical methods were used to identify the specifics of the transformation of economic relations of tourism industry subjects. The authors designed the recommendations based on the system method and modeling method.

Results

The tourism industry has shown dynamic growth over a long period of time, contributing both directly and indirectly to the creation of global GDP (Figure 4.1).

The importance of this sector of the economy is also determined by its significant overall contribution to employment (including the broader effect of investment, supply chain, and induced income). According to the World Travel & Tourism Council,

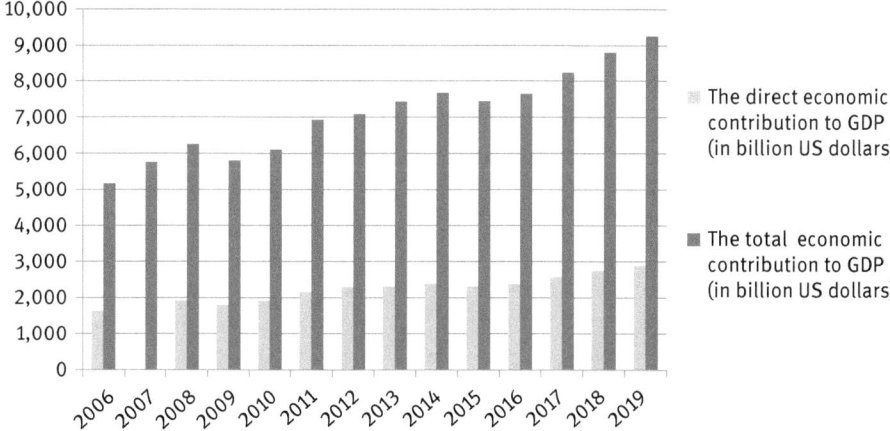

Figure 4.1: Direct and total contribution of the tourism industry to world GDP, in billions of us dollars in 2006–2019 (WWTC, 2019; UNWTO, 2017).

122,891,000 jobs were created in 2018 (3.8% of total employment) both directly in the tourism industry and in adjacent industries-the restaurant and leisure industry, directly supported by tourists (WTTC, 2019).

Intensive digitalization of the tourism industry makes global changes in its structure, and also leads to the death of certain market segments. According to the Bureau of labor statistics only between 2000 and 2014 the number of travel agencies decreased from 124,030 to 64,250 (Bureau of Labor Statistics, 2015). The number of real tourism businesses in most European countries continues to decline as a result of the huge opportunities of the Internet and due to the formation of online travel agencies (OTA). Some online travel companies show revenue growth. In 2019, Indian online travel provider MakeMyTrip had the largest revenue growth among online travel portals with a combined annual growth rate (CAGR) of 20 percent. According to GP Bullhound, MakeMyTrip's market capitalization in December 2019 was us $ 2.36 billion (Statista, 2020). In 2016–2017, Ctrip (China's largest OTA), Trivago, and Despegar (Latin America's largest OTA) also had the highest annual revenue growth rates. Other online travel companies have seen revenue growth below 5%: Tripadvisor (5%), eDreams Odigeo (4%), Lastminute (2%) (Prieto, 2018).

Digital Technologies in the System of Economic Relations of Tourism Industry Subjects

The tourism industry is an integral part of the economic system responsible for organizing the processes of creating a tourist product and bringing it to the final consumer. The subjects of the tourism industry include: manufacturers of tourist services (hotels, restaurants, transport organizations, airlines, etc.), tourist enterprises (tour operators and travel agents) and consumers who enter into economic relations regarding the production and distribution of tourist products. The structural components of the tourism industry are also the infrastructure and suprastructure of tourism, which ensure the normal and effective activities of all subjects. Traditionally, the economic relations of tourism industry subjects were based on the formation of consumer demand for a certain tourist product and its satisfaction by the manufacturer of tourist services. However, the process of promoting tourist services to the final consumer in this case did not provide for their direct connection. In the process of promotion, intermediaries appeared in the form of tourist enterprises (tour operators) that complete a tourist product from individual services, and a distribution system (travel agencies), through which the finished tourist product was brought to the consumer. However, the appearance of the first digital services in the early 2000s changed the configuration of economic relations between tourism industry subjects. Let's consider the digital technologies currently used in the tourism industry (Table 4.1).

The appearance of the described technologies and services significantly changes the traditional economic relations of the tourism industry (Figure 4.2).

The economic relations of tourism industry subjects become mediated by digital media and infrastructure, which provide them with cross-functional ownership of processes, practices and connections (Dickinson et al., 2014): 1) Manufacturers of tourist services are implementing digital technologies in order to speed up operations and expand channels for promoting services. For distribution, manufacturers now interact not only with tourist enterprises, but also directly with the consumer, differentiating their own activities and increasing the possibility of obtaining additional profit. Thus, the advent of digital platforms Booking, Aviasales opened channels of direct interaction between the manufacturer of tourist services and their consumer. 2) Tourism enterprises (tour operators and travel agents) digitalize their activities in order to increase revenue (by minimizing transaction costs associated with the search for information about new tourist services), automate document management, and expand the target audience of consumers through monitoring consumer preferences. For example, well-known brands such as Expedia, Travelocity, Priceline, or MakeMyTrip (MMT) control the global tourism market with 95% coverage in the US alone (Phocus Wright, 2017). 3) Consumers of a tourist product get an opportunity to access a wider volume of information, increase their tourist literacy, expand the boundaries of their own consumer choice when planning a tourist route that best meets their needs. According to Google statistics, during the travel planning period, consumers make about 400 search queries on the network, carefully choosing a particular route, company or hotel (Bogomazova et al., 2019).

The configuration of interactions between subjects changes, the control of tourism enterprises over consumers is decreasing, which leads to the transformation of the business environment in the tourism industry (Buhalis and Amaranggana, 2013). On the one hand, tourism enterprises add little to the tourism product compared to

Table 4.1: Digital technologies used in the tourism industry.

Digital technology	Tourism industry subject used technology	Content	Examples
Machine learning and artificial intelligence technologies	Manufacturer of services, tourist enterprise, consumer	They provide the most personalized result when planning a tourist product. They rely on information about consumer preferences and offer solutions used by other consumers. AI-systems significantly simplify the organization of a tourist trip, do not require human participation.	– The Hilton hotel chain has launched an online Concierge service. – Facebook analyzes data to offer users placement options through contextual advertising. – Booking.com introduced the Booking Experiences service, which uses AI to help consumers organize leisure activities and purchase tickets using a QR code. – Platform Booking.com accumulated data on the cost of hotel rooms and created a system for rating them by guests (Dombase, 2019). – American startup Lola created a travel bot that replaces travel agents (Finance rambler, 2017).
Internet of things technologies	Manufacturer of services, consumer	The technology is associated with personalized service, helps to collect data about the customer's preferences, and also makes it possible to optimize the environment for these parameters (temperature, noise level, lighting, water temperature).	– Smart devices set the room temperature; order room service before arrival; turn on / off lights; receive real-time flight status alerts. – Hotels install Smart look devices that remotely open and lock the door automatically or on a schedule. Smart sensors warn you about equipment wear, water or gas leaks, or the presence of damaged products in the warehouse (Dombase, 2019). – Smart tags are attached to bags at the airport and show the traveler where their personal belongings have flown.

(continued)

Table 4.1 (continued)

Digital technology	Tourism industry subject used technology	Content	Examples
Robotics technologies	Manufacturer of services, consumer	Technologies of co-bots, robots that can understand and work with people reduce the need for staff, such technology significantly simplifies the conduct of homework and hotel business.	The robot A. L. O. "Botlr" is used in hotels to deliver orders to customers, including chargers, cables for customers' appliances, snacks and morning press, it informs guests by phone about their arrival and the need to open the door to receive the delivered items (Robogeek, 2014).
Cloud computing and big data processing technologies	Manufacturer of services, tourist enterprise, consumer	Create the possibility of creating global communication platforms to improve the efficiency of service exchange, make resources liquid, increasing the density of resources and facilitating access to resource packages. (Blix, 2015).	– Operating platforms: Uber, Gett, Yandex. – Mobilisation platforms: CRM-системы, Bitrix24. – Social platforms: Facebook, Instagram. – Innovative platforms: Android, IOS, Microsoft Windows. – Integrated platforms: App Store, iCloud. – Aggregated platforms: Alibaba. – Training platforms: YouTube, Coursera.

Blockchain technologies	Manufacturer of services, consumer	A distributed registry creates a "trusted digital environment" that allows you to increase the reliability of orders, reservations, and payments by ensuring the accuracy of information and reviews about services.	– Blockchain technology is used for identification at airports, train stations, ports, during the check-in procedure, as well as when searching for Luggage. – Blockchain technology tracks the processing of documents, booking rooms and purchasing tickets, and the movement of the traveler.
Additional reality technologies	Consumer	The program visually combines two independent spaces: the world of real objects around a person and a virtual world recreated on a computer. It is formed by superimposing programmed virtual objects on top of the video signal from the camera, and becomes interactive by using special markers.	– The AR City app is used for easy navigation around an unfamiliar city. – The Airbus iflyA380 App with the AR option allows you to check the size of the legroom near the selected seat and the availability of entertainment options using your smartphone (VR-Journal, 2019). – The Air app helps customers explore flight options and choose the most suitable one, as well as compare their bag with the airline's permitted carry-on size in augmented reality (Appintheair, 2020). – Google Lens recognizes images based on artificial intelligence and translates foreign text over the original caption.

(continued)

Table 4.1 (continued)

Digital technology	Tourism industry subject used technology	Content	Examples
Virtual reality technologies	Manufacturer of services, tourist enterprise, consumer	They create the effect of full immersion due to high image quality, and the wide possibilities of displaying objects allow you to introduce the consumer to sights, including the integration of a virtual guide.	– The hotel chains Marriott, Best Western, Holiday Inn, Express & Carlson Rezidor Hotel Group have begun to implement virtual reality in their business processes by demonstrating hotel rooms and infrastructure to customers. (Dombase, 2019). – Airlines use VR to show VR presentations to customers to get them a vivid experience. – Virtual reality helmet VIVE Focus allows you to immerse customers in places and attractions that are physically inaccessible to the average tourist.

Source: compiled by the authors.

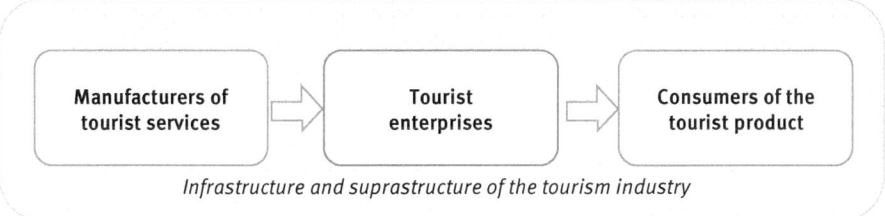

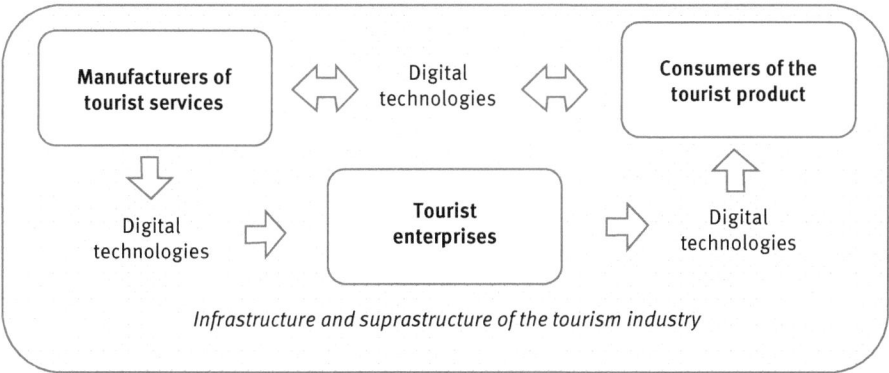

Figure 4.2: Transformation of economic relations of tourism industry subjects under the influence of digital technologies.
Source: compiled by the authors.

digital systems: they manage information and make reservations, charge additional fees that increase the overall price of tourism products, they are relatively inflexible and give less freedom of choice to consumers. On the other hand, tourism enterprises provide professional travel advice and use their expertise to save time for consumers. Professional technologies used by tourism enterprises are complex and expensive for independent use by consumers. Travel companies reduce travel insecurity and can offer better prices by buying in bulk or through consolidators, while online transactions are not yet secure or reliable (Buhalis, 1998). Thus, traditional tourist enterprises with offline offices leave the tourist market, and those companies that implement innovative business models in the best way to provide tourist services to consumers achieve competitive advantages (Cherevichko and Temyakova, 2019). For example, in 2014–2015, the sector of tourist enterprises (travel agencies) in the regions of Russia decreased by 30%, and in the period from 2015 to 2017, the number of travel agencies in the largest cities of the country decreased by another 18% (Ratanews, 2018).

Changing the Model of Creating a Tourist Product in the Conditions of Digitalization

The reason for the change in the economic relations of tourism industry subjects is the evolution of the value creation process in the conditions of digitalization. The concept of "consumer value of a product" is usually identified with the definitions "cost" and "utility". The classic approach to creating a "value chain" involves creating the consumer value of a product/service by the company for the consumer without his participation (Porter, 2005; Kotler, 2012), this approach in the digital economy gives way to approaches based on continuous interaction (Prahalad and Ramaswami, 2006). Indeed, there are more opportunities to ensure the growth of consumer value of a product in a digital environment. Both the producer and the consumer can now create value together. The so-called model of joint creation of consumer value, described in the works of Prahalad and Ramaswami (2006), Vargo and Lash (2006), Payne et al. (2008) comes into effect. An example of the practical implementation of this model is the value orchestration platform – a service system where the process of joint creation and interaction of customers and suppliers for the joint creation of new values takes place.

Prerequisites for the implementation of the model of joint creation of consumer value are: a new product is created individually.; a new product is created together with consumers; consumers' individual and social experience with resources, processes and their results becomes the core of value creation (Helkkula et al. 2012); the consumer and manufacturer are in a state of continuous collaboration in real time; the manufacturer coordinates all business processes; the consumer can compare the risks and benefits of creating a product/service in terms of information transparency; the benefits of production activities are distributed among all participants in the interaction; internal corporate consolidation and optimization of relationships with contractors and consumers (Gordon, 2014).

Such value cooperation in the tourism industry is implemented through an interactive dialogue between the consumer of a tourist product and the manufacture of tourist services (or a tourist enterprise) at each stage of the value chain. A new functional element – a community of consumers united by a digital platform for communication and exchange of experience – appears in the model of joint value creation. The consumer community now has a direct impact on the behavior of each consumer of a tourist product, and tourism enterprises lose the ability to control consumers in the process of making a purchase decision (Melis et al., 2015). The tourism industry subjects in the conditions of digitalization integrate resources in different ways. Manufactures of tourism services and tourist enterprises use operating platforms to integrate market, individual, and community resources. Consumers of a tourist product integrate individual resources and the resources of the consumer community to participate in the joint creation of values. In general, this model of creating consumer value in the tourism industry turns consumers into co-producers of the tourism product and contributes to improving their satisfaction with the final product (Figure 4.3).

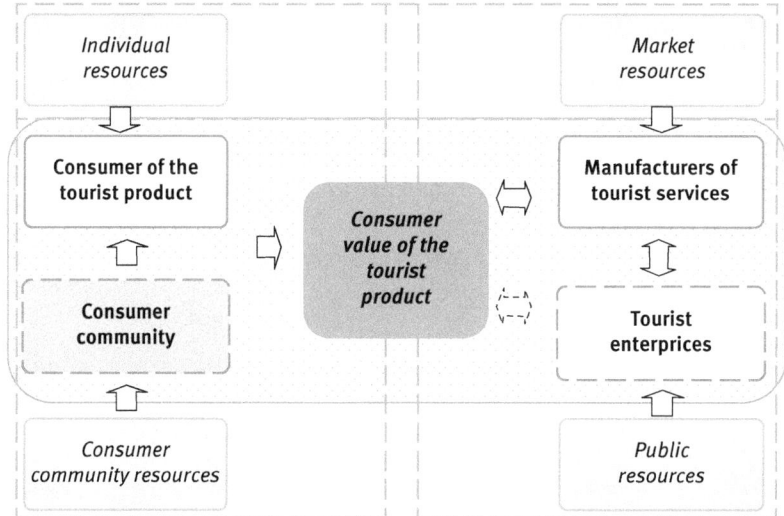

Figure 4.3: Model of joint creation of consumer value in the tourism industry
Source: compiled by the authors.

The model shows why the displacement of the part of tourism enterprises from the market in terms of digitization: 1) the growing influence of the consumer community weakens channels of the impact of tourist enterprises on the consumer; 2) strengthening the role and independence of consumers in value creation due to low adaptability of tourist enterprises; 3) low level digital equipment of tourist enterprises along with the increasing demands of the consumer to digital content.

The transition to the consumer value creation model described by Norman and Ramirez (2003) will contribute to the further transformation of economic relations between tourism industry subjects. In this case, it is likely that the manufacture of tourist services will provide resources for the process and mobilize the consumer to create value for itself, and the tourism enterprise sector will completely exhaust its potential.

Conclusion

It is necessary to review the functional load of tourism enterprises in order to maintain the configuration of economic relations of tourism industry subjects in the context of ongoing digitalization. They should move from acting as an intermediary to acting as a generator of new tourism ideas based on the use of advanced digital technologies, adapting offers for customers and covering all stages and channels of interaction with consumers. In our view, the prospects for tourist enterprises are related to the implementation of a real-time dynamic packaging system that provides a risk-free business model, the development of new booking services and online service platforms for direct sales to consumers of individual services and travel packages with the ability to independently design a tour, the use of big data-based solutions for forecasting demand and pricing management, virtualization of tourist projects, as well as maintaining an online presence and promoting services through social networks. In our opinion, the development of mobile applications for tourists, the creation of online schools for novice travelers, and the implementation of specialized service tasks are also promising.

References

App in the air (2020), "Never too early, never too late", available at: https://www.appintheair.mobi/ (accessed 25 March 2020).

Benckendorff, P., Sheldon, P. and Fesenmaier, D. (2014), Tourism Information Technology, 2nd Edition, Gutenberg Press, Mailta.

Blix, M. (2015) The Economy and Digitalization: Opportunities and Challenges, Confederation of Swedish Enterprise.

Bogomazova, I., Anoprieva, E. and Klimova, T. (2019), "Cifrovaya ekonomika v industrii turizma i gostepriimstva: tendencii i perspektivy", Servis v Rossii i za rubezhom, Vol. 13, No. 3 (85), pp. 34–47. doi: 10.24411/1995-042X-2019-10303.

Buhalis, D. (1998), "Strategic Use of Information Technologies in the Tourism Industry", Tourism Management, Vol 19 (5),pp. 409–421. DOI: 10.1016/S0261-5177(98)00038-7.

Buhalis, D. and Amaranggana, A. (2013), Smart tourism destinations. Information and Communication Technologies in Tourism. Springer International Publishing.

Bureau of Labor Statistics (2015), "The number of freelance travel agents went from 124,030 in 2000 to 64,250 in 2014", available at: https://www.nomadicmatt.com/travel-blogs/using-travel-agents/ (accessed 30 March 2020).

Cherevichko, T. and Temyakova, T. (2019), "Cifrovizaciya turizma: formy proyavleniya", Izv. Sarat. un-ta. Nov. ser. Ser. Ekonomika. Upravlenie. Pravo, Vol. 19 (1), ss. 59–64.

Coates, J. (1993), "The future of tourism: the effect of science and technology", Vital Speeches, Vol, 58(24), pp. 759–763.

Dickinson, J., Ghali, K., Cherrett, T., Speed, C., Davies. N. and Norgate, S. (2014), "Tourism and the smartphone app: capabilities, emerging practice and scope in the travel domain", Current Issues in Tourism, Vol. 17 (1), pp. 84–101.

Dombase (2019), "Umnye chemodany, virtual'nye tury i IoT dlya bezopasnosti turistov", available at: https://rb.ru/opinion/tehnologii-v-turizme/ (accessed 25 January 2020).

Finance.rambler (2017), "Iskusstvennyj intellekt v trevel-industrii", available at: https://finance.rambler.ru/economics/37251846/?utm_content=finance_media&utm_medium=read_more&utm_source=copylink (accessed 17 March 2020).

Gordon, Y. (2014), Marketing partnyorskih otnoshenij, Piter, Sankt-Peterburg.

Gyr, U. (2010), "The History of Tourism: Structures on the Path to Modernity", in: European History Online (EGO), published by the Institute of European History (IEG), Mainz 2010-12-03, available at: http://www.ieg-ego.eu/gyru-2010-en (accessed 10 March 2020).

Helkkula, A., Kelleher, C. and Pihlstrom, M. (2012), "Characterizing value as an experience: implications for service researchers and managers", Journal of Service Research, Vol. 15(1), pp. 59–75.

Kotler, F. (2012), Marketing menedzhment, Piter, Sankt-Petersburg.

MacKay, K. and Vogt, C. (2012), "Information technology in everyday and vacation contexts", Annals of Tourism Research, Vol. 39 (3), pp. 1380–1401.

Melis, G., McCabe, S. and Del Chiappa, G. (2015), "Conceptualizing the Value Co-Creation Challenge for Tourist Destinations: A Supply-Side Perspective", Marketing Places and Spaces (Advances in Culture, Tourism and Hospitality Research, Vol. 10), Emerald Group Publishing Limited, pp. 75–89.

Murdaugh, M. (2005). Fundamentals of Destination Management and Marketing, American Hotel & Lodging Educational Institute, Michigan.

Norman, R. and Ramirez, R. (2003), "From value chain to value constellation: designing interactive strategy", Harvard Business Review, July-August, pp. 65–77.

Payne, A., Storbacka, K. and Frow, P. (2008), "Managing the co-creation of value", Journal of the Academy of Marketing Science, No. 36. pp. 83–96.

Phocus Wright (2017), "U.S. Consumer Travel Report Ninth Edition (Series)", available at: https://www.phocuswright.com/Travel-Research/Consumer-Trends/U-S-Consumer-Travel-Report-Ninth-Edition-Series (accessed 20 January 2020).

Poon, A. (1993), Tourism, technology and competitive strategies, CAB International, Oxford.

Porter, M. (2005), Konkurenciya, Izd. dom Vil'yams, Moskva.

Prahalad, K. and Ramasvami, V. (2006), Budushchee konkurencii. Sozdanie unikal'noj cennosti vmeste s potrebitelyami, Olimp-Biznes, Moskva.

Prieto (2018), "10 Online Travel Public Companies – Full Year 2017 Results", available at: https://medium.com/traveltechmedia/10-online-travel-public-companies-dee6df73f768 (accessed 12 March 2020).

Ratanews (2018), "Ofisy turagentstv v krupnejshih gorodah", available at: http://ratanews.ru/i/editor_upload/images/untitled-2(153).jpg (accessed 20 January 2020).

Robogeek (2014), "Servisnye roboty", available at: http://www.robogeek.ru/servisnye-roboty/robotov-nachali-ispolzovat-v-gostinichnom-servise (accessed 11 March 2020).

Statista (2020), "Leading online travel companies worldwide 2019-2021, by CAGR", available at: https://www.statista.com/statistics/1039631/leading-online-travel-companies-by-cagr/#statisticContainer (accessed 15 March 2020).

Tohology (2020), "Po dannym Vsemirnoj turistskoj organizacii, v 2019 godu v mire zaregistrirovano 1,5 milliarda mezhdunarodnyh turisticheskih poezdok", available at: https://www.tohology.com/news/tags/?>f=other_tags&t=Всемирная%20туристская%20организация (accessed 10 March 2020).

UNWTO (2017), "Tourism Highlights. 2017 Edition", available at: http://tourlib.net/wto/WTO_highlights_2017.pdf (accessed 30 March 2020).

Vargo, S. and Lash, R. (2006), "Razvitie novoj dominiruyushchej logiki marketinga", Rossijskij zhurnal menedzhmenta, Vol. 4, № 2, pp. 73–106.

VR-Journal (2019), "Ispol'zovanie dopolnennoj real'nosti v turizme", available at: https://vr-j.ru/news/3-primeneniya-dopolnennoj-realnosti-v-industrii-turizma/ (accessed 15 March 2020).

World Travel & Tourism Council (WTTC) (2019), "Travel & Tourism: Economic Impact 2019", available at: https://www.wttc.org/-/media/files/reports/economic-impact-research/regions-2019/world2019.pdf (accessed 12 February 2020).

Xiang, Z. and Gretzel, U. (2010), "Role of social media in online travel information search", Tourism Management, Vol.31 (2), pp. 179–188.

Part II: **Finances, Corporate Accounting and Management in Industry 4.0**

Irina G. Sergeeva, Irina E. Zuber, Vera D. Nikiforova
and Alexander A. Nikiforov

5 Russian Foreign Loans Market: Problems of Balancing

Introduction

In the last few years, active discussion on balancing markets was caused by the global financial crisis that affected both developed countries and those with developing market infrastructure. One of most common impacts was turning the financial crisis into sovereign debt crisis. Many countries had to reevaluate credit risks to estimate the probability of sovereign bond default. Countries with comparable (before the crisis) premiums for sovereign risk suddenly showed essential inequality and unpredictable vicissitudes in the balance between interest rates and credit risk. For players in private or public sectors, true and reliable information on premiums for sovereign risks helps to assume the risks and volatility of the market. At the present time, there are many works studying the nature of the premium for sovereign risk and factors explaining its dynamics. Thus, some authors (Fontana and Scheicher, 2010) found that standard factors influence on the premium, e.g. changing investors' appetite for risk. Studies (Caceres and Unsal, 2011) of major Asian countries has showed that the main ground for changing sovereign spreads is external factors affecting the sovereign risk.

Another research worker (Santis, 2012) studied eurozone countries, having found three factors affecting sovereign spreads, in particular: overall risk factor in the region in general; credit risk for a particular country; negative impact of situation in eurozone countries, e.g. debt problems in Greece or Spain.

Cost of debt of developing countries on the world financial market is defined by EMBI (Emerging Markets Bond Index), given by J.P. Morgan Chase Bank. Obviously, when the movement of spreads depends on internal macroeconomic factors, it means that a stable macroeconomic policy is implemented. Such policy reduces default risk, and, hence, the spread. Essential rise of EMBI spread (i.e. rise of default risk) caused

---

Note: The chapter was prepared in the framework of fundamental research in the field of regulation of the financial market in increasing aggravation of risks.

Irina G. Sergeeva, ITMO University, Saint Petersburg, Russia
Irina E. Zuber, Institute of Mechanical Engineering of the Russian Academy of Sciences, Saint Petersburg, Russia
Vera D. Nikiforova, Alexander A. Nikiforov, Financial University under the Government of the Russian Federation (St. Petersburg Branch), Moscow, Russia

https://doi.org/10.1515/9783110654486-005

by unfavorable change of market situation in developed countries may raise the ratio of government debt in foreign currency to GDP in developing countries.

In case when this disbalance of foreign currency debt takes place together with devaluation of a national currency and rising internal interest rates, crisis developments may take place in developing countries economies, even if internal macroeconomic situation is stable.

Methodology

The problem of market stabilization is considered. We have constructed mathematical model of considered system as nonlinear indeterminate system of differential equations in matrix form. In such systems coefficients of matrix are functionals of arbitrary nature but what is known about these functionals is the boundaries of variation only.

Using bond index for emerging markets we can analyze how the growth of countries' credit spreads affect the government bond market of a particular country. During financial crises credit spreads on emerging markets may rise consistently as a result of interaction between change in spreads and volatility.

Spreads instability in unfavorable market situation witnesses the negative impact of financial globalization to world government bond markets, but especially to emerging ones. Some papers investigate the causes of high growth of credit spreads on such markets. Furthermore, a number of authors (Baek et al., 2005, Diaz-Weigel and Gemmill, 2006, Eichengreen and Mody, 1998, Remolona et al., 2008) tried to explain why this growth promotes market instability, this fact, in its turn, leads to depression of a national market. These authors take the spread growth as a result of changes in market sensitiveness and in risk premiums. Though these are really important factors leading to market instability, they do not explain why spreads then continue to keep up soaring. Soaring spreads (a result of unpredictable external turmoil) may lead to a crucial financial crisis in these countries. This is one of major problems for emerging government bond markets.

The lower the spread or volatility are, the better conditions for external loans. Generally, the main factors affecting the spread are as follows: a unique credit risk due to country specifics; global risks common for all countries.

Sometimes the spread can deviate from the level determined by these factors. These market deviations are asymmetrical, since in the period of spread growth they are larger than during spreads reducing. Especially, strong asymmetric volatility is typical for emerging government bond markets. That is why spreads on emerging markets may grow, even when there is no evidence of change in fundamental factors.

Many authors try to explain the causes of such deviations. Some of the researchers consider that the change of spreads depends on global factors, rather than country-specific ones (Ciarlone et al., 2009, Kamin and Kleist, 1999, Longstaff et al., 2007, McGuire and Schrijvers, 2003, Fedorova and Pankratov, 2010, Sergeeva, 2012). Others explain essential spread change during financial crises as a result of behavior of market participants. They assume that time-varying market sensitiveness (Eichengreen and Mody, 1998), market psychology and risk premium (Remolona, et al., 2008) affect participants' adversion. These studies note that growing risk premium may burst of the spread, but they do not explain why this process still is going on for some uncertain period of time.

The major problem of spread variations on emerging government bonds markets is their ability to burst the market stability in case of a severe spread change. Assuming a positive link between volatility and risk premium, it should be noted that spread soaring is going on together with to time-varying dynamic interaction between spread change and volatility. The participants' responsibility for volatility growth is determined by information they possess. If they are well informed, change in bond pricing will be small, volatility low, and all this will allow the market to come back to balance in a relatively short time. If they are not informed well enough, their actions promote change in bond prices, and thus raise volatility. Thus, lack of information can make the financial market inefficient, leading to periodical changes in bond prices preventing the market from coming back to balance.

Results

Now we will formulate and solve the task of stabilizing state external loans market and will formulate corresponding mathematical model.

Consider three variables:
- x_1 – deviation of volatility from the desirable condition,
- x_2 – deviation of spread from the desirable condition, and
- x_3 – deviation of risk from the desirable condition and form vector of deviations.

Then the model of the market with some lack of information takes a following form

$$\dot{x} = A(x,t)x + M(\cdot)x + bu \qquad (1)$$

where $\dot{x} = \frac{dx}{dt}$, $A(x,t) - (3\times 3)$- matrix with fully defined bonded coefficients, $M(\cdot)$ is uncertain matrix. It is known only bond of Euclidean norm of $M(\cdot)$ designed usually as $|M|$, i.e.

$$|M| < \mu \qquad (2)$$

Assuming that only x_3 is influenced we have v

$$b = \begin{vmatrix} 0 \\ 0 \\ 1 \end{vmatrix}$$

where

$$\det| b \ A(x,t)b \ A^2(x,t)b | \geq 0 \qquad (3)$$

Consider the stabilizing control for (1)–(3) in form

$$u(x,t) = S^T(x,t) \cdot x \qquad (4)$$

Here and later

$[^T]$ – sign of junction; $||$ – sign of Euclidean norm.

So closed-loop system (1)–(4) has a form

$$\dot{x} = (Q(x,t) + M(\cdot)) \cdot x \qquad (5)$$

$$Q(x,t) = A(x,t) + b \cdot S^T(x,\ t) \qquad (6)$$

This system (5) (6) is nonlinear, nonstationary and uncertain. So, we must define the stabilizing control (4) that provides existence of the Lyapunov function for this system (Zuber and Gelig, 2012, Gelig et al., 2006).

Consider the Lyapunov function for (5) (6) as

$$V(x) = x^T \cdot x \qquad (7)$$

Then

$$\dot{V}(x) = x^T L(x,t) \qquad (8)$$

where

$$L(x,t) = Q^T(x,t) + Q(x,t) + M^T(\cdot) + M(\cdot) \qquad (9)$$

So our goal is to find vector $s(x,t)$ that for any number $\delta > 0$ provides the fulfillment of condition

$$L(x,t) < -\sigma I$$

With the account to (3) we have a spectral decomposition of matrix (6) [14, 15].

$$Q(x,t) = \sum_{i=1}^{3} \lambda_i d_i(x,t) g_i(x,t)$$

5 Russian Foreign Loans Market: Problems of Balancing

where $\lambda_i < 0$ $i = 1, 2, 3$ are spectrum of $Q(x, t)$

$$d_i(x, t) = (A(x, t) - \lambda_i I)^{-1} b, \tag{10}$$

vectors $g_i(x, t)$ are defined by formula

$$g_i^T(x, t) \; d_j(x, t) = \begin{cases} 1 & i = j \\ 0 & i \neq j \end{cases}. \tag{11}$$

Then vector $s(x,t)$ is defined by system

$$s^T(x, t) d_i(x, t) = -1. \tag{12}$$

Assume the following choice

$$\lambda_i = -\lambda - \tau_i \;\; \tau_i \neq \tau_j \;\; i \neq j \;\; 0 < \tau_i < \tau \;\; i = 1, 2, 3.. \tag{13}$$

$0 < \lambda$ – the big parameter (or arbitrary big number).

Consider the order relation for function $F(\lambda)$ and number λ and design it as $O_\lambda(F)$. Accounting to (10)

$$O_\lambda(d_i(x, t)) = -1 \tag{14}$$

Later we drop the arguments in following formulas for simplicity and consider matrix

$$D = \sum_{i=1}^{3} d_i \cdot d_i^T \tag{15}$$

With account to (14)

$$O_\lambda(D_0) = 0 \quad \text{where} \quad D_0 = \frac{D}{|D|}$$

$$O_\lambda(|D_0|) = 0 \quad O_\lambda(|D_0^{-1}|) = 0 \tag{16}$$

Consider matrix

$$P = DLD \tag{17}$$

As $D = D^T$, then sign P = sign L

With account to (11)

$$P = -2 \cdot \lambda \cdot D^2 - N + D(M^T(\cdot) + M(\cdot))D \tag{18}$$

Where

$$N = D \sum_{i=1}^{3} \tau_i \cdot d_i \cdot d_i^T - \sum_{j=1}^{3} \tau_j \cdot d_j \cdot d_j^T D \tag{19}$$

So $|N| \leq 2\tau\ |D|2$ and in account to (17)

$$L \leq -2\lambda I + 2\tau|D_0^{-1}|^2 I + 2\mu I \qquad (20)$$

So, choice λ and τ from inequality

$$-\lambda - \tau|D_0^{-1}|^2 + \mu < -\sigma \qquad (21)$$

with account to (16) fulfills the synthesis of stabilizing control for closed-loop system (5) (6).

Conclusion

On the base of the results coming from the paper the following summary can be highlighted.

Investigating and analyzing interdependence between risk, volatility and sovereign credit spread, the authors highlight the determinants explaining the nature of market instability: (1) impact of external factors, especially for countries with emerging financial markets; (2) impact of country-specific factors with negative effect on credit rating and sovereign credit spreads; (3) insufficient information transparency of the market and lack of information for the participants.

Countries with emerging financial markets should create and develop stabilization mechanism facilitating efficiency of obtaining sufficient information on the government bond market. As a result of the study, the functioning model for the Russian external borrowings market is considered in the form of nonlinear and non-stationary control system for which the solution of stabilization problem is obtained.

References

Baek, I.M. et al. (2005), Determinants of Market Assessed Sovereign Risk: Economic Fundamentals or Market Risk Appetite? Journal of International Money and Finance, Vol. 24 (4), pp. 533–548.
Caceres, C. and Unsal, D.F. (2011), Sovereign Spreads and Contagion Risks in Asia, IMF Working Paper, No. 11/134.
Ciarlone, A. et al. (2009), Emerging Markets Spreads and Global Financial Conditions, Journal of International Financial Markets, Institutions and Money, Vol. 19 (2), pp. 222–239.
Diaz-Weigel, D. and Gemmill, G. (2006), What Drives Credit Risk in Emerging Markets? The Role of Country Fundamentals and Market Co-Movement, Journal of International Money and Finance, Vol. 25(3), pp. 476–502.
Eichengreen, B. and Mody, A. (1998), What Explains Changing Spreads on Emerging-Market Debt: Fundamentals or Market Sentiment? Working Paper, National Bureau of Economic Research, Cambridge, MA, No. 6408.

Fedorova, E.A. and Pankratov, K.A. (2010), The Impact of Macroeconomic Factors on the Russian Stock Market, Studies on Russian Economic Development, №2.
Fontana, A. and Scheicher, M. (2010), An Analysis of Euro Area Sovereign CDS and Their Relation with Government Bonds, ECB Working Paper, No. 1271.
Gelig, A. Kh. et al. (2006), Stability and Stabilization of Nonlinear Systems, St. Petersburg University.
Kamin, S.B. and Kleist, K. Von. (1999), The Evolution and Determinants of Emerging Market Credit Spreads in the 1990s, Working Paper, Bank for International Settlements, Basel, No. 68.
Longstaff F.A. et al. (2007), How Sovereign Is Sovereign Credit Risk? Working Paper, National Bureau of Economic Research, Cambridge, MA, No. 13658.
McGuire, P. and Schrijvers, M. (2003), Common Factors in Emerging Market Spreads, BIS Quarterly Review (December).
Remolona, E. et al. (2008), The Dynamic Pricing of Sovereign Risk in Emerging Markets: Fundamentals and Risk Aversion, Journal of Fixed Income, No. 17, pp. 57–71.
Santis, R.A. (2012), The Euro Area Sovereign Debt Crisis Safe Haven, Credit Rating Agencies and the Spread of the Fever from Greece, Ireland and Portugal, ECB Working Paper, No. 1419.
Sergeeva, I.G. (2012), Financial-Real Sector Interactions in the Post Crisis World", Scientific journal of ITMO University. Series of Economics and Environmental Management, No 1.
Zuber, I.E. and Gelig, A. Kh. (2012), Using the direct and indirect control to stabilize some classes of uncertain systems, Automation and remote control, Vol. 73, No. 8.

Elena G. Patrusheva, Elena I. Lifanova and Anna V. Raikhlina

6 The Problem of Monitoring the Effectiveness of Projects in the Process of their Implementation

Introduction

In the public consciousness the fourth manufacturing revolution is associated with radical technologies, fully automated digital production, creation of global industrial networks, etc. All these innovations are being implemented through the project management (PM) approach. At the same time, high level of novelty in such projects create emerging risks for their clients which require updating the methodological support for their implementation in terms of customer satisfaction.

In PM literature and PM international and national standards project processes are described with a clear distinction between them and operations. Project managers are expected to manage the triple constraint of time, cost and scope/quality while the project customer, focusing on achieving certain benefits, expects project effectiveness (business outcomes) from using the project product in operations. And, if so, some discrepancy between the project team's aims and the customer's expectations may be detected. Thus, the development of methodological and organizational tools for regular monitoring the effectiveness of projects in the process of their implementation is in demand.

Methodology

The project management standards and guidelines have three interrelated levels: international, national and professional communities' one. International standards include the ISO 21500:2012 Guidance on Portfolio Management; A Guidebook of Project and Program Management for Enterprise Innovation; Project Manager Competency Development Framework (PMCDF), etc. National standards are based on international experience in consideration of the country's specificities. For example, in Russia, project activities are regulated by GOST R 54869 – 2011; GOST R 54870 – 2011; GOST R ISO 21500 – 2014, etc. According to the standards the interdependency between project process groups requires the controlling process group. The

---

Elena G. Patrusheva, Elena I. Lifanova, P.G. Demidov Yaroslavl State University, Yaroslavl, Russia
Anna V. Raikhlina, Financial University under the Government of the Russian Federation, Yaroslavl, Russia

https://doi.org/10.1515/9783110654486-006

controlling processes are used to monitor, measure and control project performance against the project plan (ISO, 2012).

Monitoring procedures in terms of frequency, timing, and documentation are emphasized in works by Lipke, U. (Lipke, 2013), Lukina, I. A. (Lukina et al., 2017), Simonov, A. V. (Simonov, 2018). In some works, emphasizing project costs monitoring (Videman, 2007), (Maslovsky, 2012), (Khotelnikhov, 2016), the method of work completed is described. A number of approaches have been used to examine project risks monitoring and evaluation (Dorokhina, 2009), (Eckles et al., 2014). Certain quality requirements are established for project products, and for these requirements some papers describe monitoring procedures throughout the project life cycle (Popova, 2016), (Duggal, 2010). Ensuring the interactions between project team and other project participants, i.e. project communications monitoring, is also reflected in researches (Dashkov, 2018). Thus, extensive literature within these fields focuses on specific subject groups monitoring. A more comprehensive approach to monitoring as one of controlling processes has been described in a few papers (Patrusheva, 2015), (Guseva, 2014). At the same time in the researches dedicated to the project effectiveness evaluation (Gritsenko, 2009), (Ruchlina, 2015), (Gussack, 2015) the link between project effectiveness an PM is absent.

Results

In this exploratory study, Net Present Value (NPV) as the primary measurement tool for project effectiveness will be used. The proposed NPV calculation model includes indicators reflected the results of both PM and operations.

$$NPV = -\sum_{n=0}^{N}\frac{C_n}{(1+WACC)^n} + \sum_{n=N+1}^{M}\frac{P_n}{(1+WACC)^n} \qquad (1)$$

The proposed model consists of basic scheduling constraints plus capital restrictions in which C_n is the cost of project works for the period n; N is the quantity of time periods in the total duration of the project; M is the quantity of time periods in the total duration of the operations; P_n is operations results for the period n and $WACC$ is weighted average cost of capital.

From Model (1) it is clear that the project NPV is determined with the duration of projects (N), their cost (C) and WACC. Thus, time, cost and financing conditions should be considered as the key constraints for the project effectiveness. Consequently, it is their changes that are evaluated during the project effectiveness monitoring.

Model (1) shows that NPV can decrease:
a. with increasing N, because it will decrease M. This circumstance could have a particularly adverse effect for innovative projects implemented on a market segment that is at stage of "skimming the cream off"

b. with increasing C, because it will course additional investment cost (especially "expensive" if compared to deferred revenues taking into account the time value of money)
c. with increasing C due to additional investment at the expense of own capital (more "expensive" if compared to loan capital), which will increase WACC

To satisfy stakeholders and deliver business, project managers need to broaden their perspective to include the following:

1. Preventive and corrective actions may be taken by project managers, when necessary, in order to achieve project objectives. If there are no changes of the key project constraints (N, C, WACC) project effectiveness monitoring is not in demand. Otherwise there is a need in assessment of their deviations negative impact.
2. NPV does not reflect the level of effectiveness. To more accurately measure the internal rate of return (IRR) calculation can be used. The difference (IRR – WACC), i.e. yield spread, is necessary to evaluate the level of project effectiveness. The customer analyzing risks both in PM and in further operations, can substantiate the minimum yield spread.
3. Change requests are made while keeping within the level of project effectiveness. To select the optimal project change direction the highest yield spread is used. The algorithm of project effectiveness monitoring organization is defined in Figure 6.1.

    Figure 6.2 elaborates on Figure 6.1 to show the interactions among the processes inside project effectiveness monitoring.
4. Selecting the optimal project change direction, the impact of the key project constraints deviations on the project effectiveness is explored.
5. If the negative impact of the key project constraints (N, C, WACC) deviations on the project effectiveness is supposed, the customer or his representatives should be involved in the project effectiveness monitoring.
6. For some organizational projects, the objective may be, for example, the achievement of certain operational results when using projected transformations in operations. In this case the project managers should consider achievement of a certain level of operations effectiveness as the project objective and monitor it the with the customer or his representatives. Herewith the project managers must have the appropriate competencies.
7. When implementing programs that include several projects focused on specific objectives, the PM standards offer to assess the costs of a certain project and its outcome for the program. In addition, the closing point of the program can be set at the operations stage, for example, after the cost recovery. In this case the program managers should monitor the project effectiveness and include an operations manager in the project team. The general algorithm for taking decisions in project change management is like the one described earlier, but the project managers have the primary responsibility for the project effectiveness monitoring.

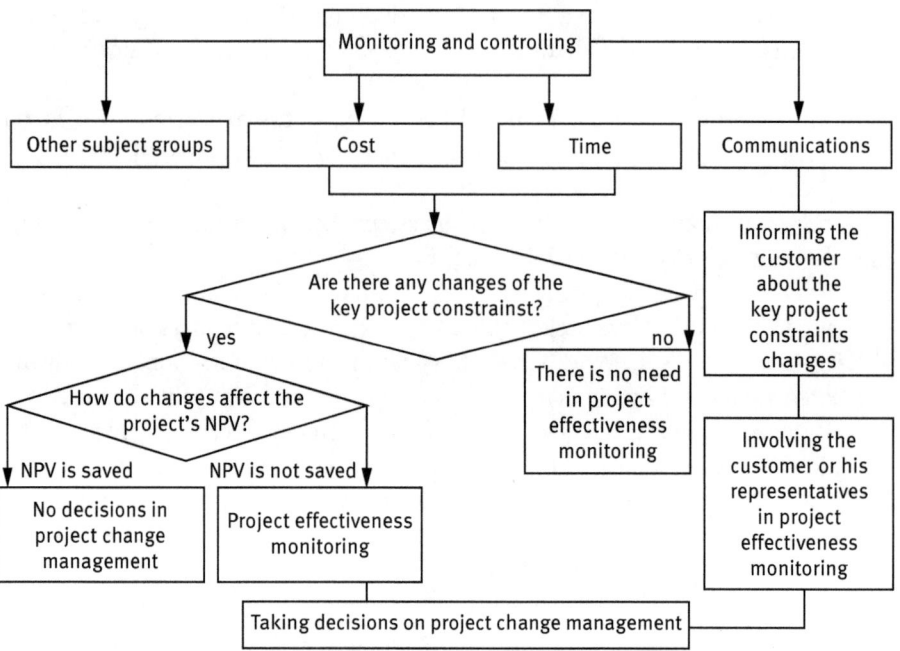

Figure 6.1: Algorithm of project effectiveness monitoring organization.
Source: developed and compiled by the authors.

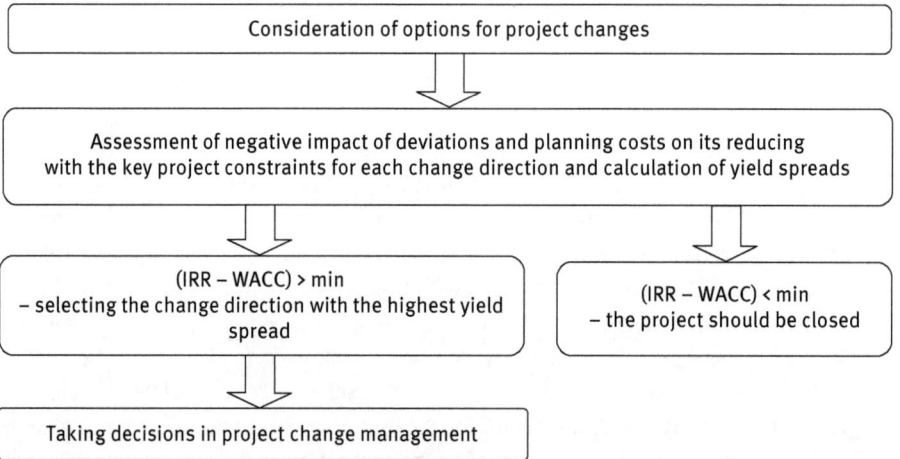

Figure 6.2: Steps of project effectiveness monitoring process.
Source: developed and compiled by the authors.

Conclusion

When implementing complex innovative projects typical of the fourth manufacturing revolution with a high probability of deviation of its parameters from the planned level, project effectiveness monitoring is needed, because the project effectiveness is the customer's objective as a filter to determine projects implementation. The controlling process group is used to monitor and measure project performance against the project plan. If the negative impact of the key project constraints (N, C, WACC) deviations on the project effectiveness is supposed, the project effectiveness monitoring should be conducted. And the decision-taking procedures for changes should be made according to the criterion of the project effectiveness maximum preservation.

In projects that imply transfer the finished product to the customer and closing before the operations start, the project effectiveness monitoring, taking into account all the key project constraints changes that occurred during its implementation, belongs to the customer's interests and activities, and must be performed by the customer or his representatives to take the right decisions in project change management.

When the project closing condition for the customer organization is to achieve a certain level of operational results, and the project implementation includes an operational stage, or has proceeded in parallel with the operations, the project managers should be responsible for the project effectiveness. Herewith the project managers must have the appropriate competencies.

Adoption the proposed procedures of project effectiveness monitoring will contribute to reach a consensus between the customer' and the project team's interests. And monitoring the effectiveness of projects in the process of their implementation should be considered as one of the PM challenges in in the current context.

References

A Guidebook of Project and Program Management for Enterprise Innovation available at: https://pmpractice.ru/knowledgebase/normative/projectstandarts/p2m/ (accessed: 30.01.2020).

Dorokhina, E.U. (2009) Project risk management: methods and tools. Environmental Economics. 2, pp. 67–80.

Duggal, J. S. (2010) How Do You Measure Project Success? Rethinking the Triple Constraint. Community Post, available at: https://projectize.com/tripleConstraint.html (accessed: 20.12.2019).

Dushkov, R. U., Tislenko, A. V. (2018) System of monitoring and controlling the project stakeholders' activities based on the method of managing duration. MID (Modernization. Innovations. Development). vol. 9. 1, pp. 86–97.

Eckles, D., Hoyt, R., Miller, S. (2014) The impact of enterprise risk management on the marginal cost of reducing risk: Evidence from the insurance industry. Journal of Banking & Finance. 43, pp. 247–261.

Federal Agency on Technical Regulating and Metrology, available at: https://www.gost.ru/portal/eng/home/standards/catalogue (accessed: 03.02.2020).

Gritsenko, D. V. (2009) Evaluation of the financial efficiency of innovation and investment projects. Economics and Efficiency of Production Organization. 11, pp. 237–240.

Guseva, E. G. (2014) Project monitoring and evaluation. CRNO, St. Petersburg.

Gusack, U. S. (2015) Project effectiveness evaluation. Bulletin of the Moscow international higher business school MIRBIS. 1, pp. 80–83.

ISO 21504:2012 "Guidance on portfolio management", available at: https://www.iso.org/standard/61518.html (accessed: 12.01.2020).

Khotelnikhov, I. K. (2016) Using the method of work completed in PM. International Scientific Review. 11 (21), pp. 62–64.

Lipke, U. (2013). Method of meeting deadlines: contribution to project management. Project and Program Management. 2, pp. 86–101.

Lukina, I. A., Gulina, M. S., Tomilko, V. A., Varosyan, A. M. The system of monitoring and controlling the progress of the project. Innovative research: theory, methodology, practice 2017 proceedings of the international conference in Penza, Russia, 2017, Nauka i Prosvetshenie, Penza, pp. 51–53.

Maslovsky, V.P. (2012) Methods of project cost monitoring. Management in a modern organization: experiences, challenges and prospects 2012 proceedings of the Vth International scientific and practical conference in St. Petersburg, Russia, 2018, Azbuka, St. Petersburg, pp. 516–522.

Project Manager Competency Development Framework, available at: https://www.pmi.org/learning/library/project-manager-competency-development-framework-7376 (accessed: 12.01.2020).

Popova, O. V. (2016) Monitoring indicators of project quality management. Bulletin of Omsk University. Series: Economy. 4, pp. 110–117.

Putrusheva, E. G, Lifanova, E. I. (2015) Methodical instruments for innovation projects implementation stages results diagnostics. Economics and Entrepreneurship. 8–1 (61–1), pp. 888–892.

Ruchlina, E. R. (2015) The review of project evaluation techniques. Current Problems of Humanities and Natural Sciences. 6–1, pp. 213–220.

Simonov, A. V., Glebova, O. V. Formation of requirements for domestic project management systems. Digital economy and industry 4.0: new challenges 2018 proceedings of the scientific and practical conference with international participation in St. Petersburg, Russia, 2018, Peter the Great St. Petersburg Polytechnic University, St. Petersburg, pp. 441–447.

Videman, R. M. (2007) Monitoring of project costs: It is easy. Project Management. 1, pp. 4–9.

Nodira B. Abdusalomova and Zarina U. Tashkenbaeva
7 The Use of Digital Technologies in the Implementation of Internal Control in the Management Accounting System

Introduction

According to the World Steel Association (WSA), steel production in March 2020 was 147.1 million tons in 64 countries, which currently supply raw materials and produce finished metal products in the global iron and steel industry. This indicator has decreased by 6.0% compared to March of 2019, which was due to ongoing difficulties caused by the Covid-19 pandemic. World steel production has been generally increasing by 1–5% compared to the previous year.

World steel production for the first three months of 2020 was 443.0 million tons, which is 1.4% less than for the same period of 2019. Asian countries produced 315.2 million tons of steel in the first quarter of 2020, which is 0.3% less than in the first quarter. European Union produced 38.3 million tons of steel in the first quarter of 2020, which is 10.0% less than in the same quarter in 2019. Steel production in North America for the first three months of 2020 was 29.5 million tons, which is 4.0% less than for the first quarter of 2019.

Iron and steel industry is one of the key sectors of the modern world economic system. The level of activity of other key industries largely depends on the level of development of iron and steel industry. Today, the importance of iron and steel industry makes itself evident in the fact that solution of internal problems in the industry is of paramount importance not only within a separate enterprise, but also within the region.

The author has shown the complexity of production processes of the major iron and steel enterprises and the continuity of the engineering process used in various places and systems, depending on the efficiency of control over the operation of objects and measures of internal control.

At the same time, qualified operators are required for industrial process management during internal control in the process of manufacture of products; each of them has an impact on the finished-product output. This necessitates the creation and automation of the algorithm of a complex internal control system for the creation of an automated internal control system, the interrelations between input and output values, and the effect.

Nodira B. Abdusalomova, Zarina U. Tashkenbaeva, Tashkent State University of Economics, Tashkent, Uzbekistan

https://doi.org/10.1515/9783110654486-007

The organization of management and internal control with the use of automation of key indicators in the iron and steel enterprises is much more complex. This is due to high temperature, chemical compounds and processing costs associated with the product manufacture process.

Materials and Method

The general scientific research principles, based on data collection, monitoring, study of normative legal documents that provide a general and systemic approach to the study of economic relations and events from the standpoint of their development and interdependence, served as the methodological basis for the research. The author of the research used logical, general and systemic approaches to general scientific and special research methods – analysis and methods for assessing economic phenomena. The systemic approach was used for the refinement of the concept of internal control, analysis and assessment of the concept of the formation of internal control system, management accounting system, and internal control system. The general approach to the problem consisted in the pursuance of the research on the organization of internal control, management accounting, economics of the iron and steel industry, and management of the iron and steel industry.

Results

The problems of internal control and accounting of manufacturing costs, as well as development of management accounting systems were examined by several foreign and domestic scientists.

Foreign economist S. Sheviakova has turned her attention to internal control in this matter and recognized that "internal control is a control activity which is regulated by in-house regulations of an enterprise, assessment of various aspects of activity and consulting of all management levels by special departments" (Sheviakova, 2017). We are of the opinion that the definition of internal control that was given by S. Sheviakova reveals the true identity of internal control. Within this meaning, the essence of internal control is oriented towards not only the control function, but also towards indicators which assess the operating efficiency of an enterprise. That is, according to this approach, the concept of internal control and its essence are supplemented with the management and assessment function. According to the economist L. Paramonova, one of the business management functions is the monitoring of accounting objects and processes with a view to checking compliance with existing laws and regulations (Paramonova, 2012). However, his point of view is narrow, as it is focused on the monitoring problem only.

In particular, according to L. Voronina, control should be regarded as the function of independent management, that is, as a separate kind of activity, which includes purposefulness, content and methods of verification (Voronina, 2014). According to the scientist, control is a well-targeted method. However, we can see that the primary function of internal control cannot be developed in this matter.

The internal control system can serve as a separate enterprise management function; alternatively, internal control can be manifested as a connection with the management account, as it could be observed in the recent theory and practice. This enables the management accounting system to acquire additional information, and the internal control system allows applying new methods (Abdusalomova, 2019).

Parts of internal control can be defined as follows: control environment; instruments of process control and management; management entity and the information system (Abdusalomova, 2017a; Abdusalomova, 2017b).

XBRL reporting which integrates not only financial information, but also strategic and operational indicators of business performance, serves as the informational basis in the context of the rapid development of digital technologies in management for the organization of the internal control system. The team of Russian scientists (Astaf'eva, O.V. et al., 2020) are of the opinion that information support of the modern internal control system must be based on XBRL reporting.

In addition, many scientists are of the opinion that the integration of preventive internal control of business processes into the corporate system, in other words, control that is aimed at ensuring observation of current regulations, procedures, and regulatory documents specifying the procedure for the presentation of facts of economic activity of a company in the corporate accounting system, is a current trend in business management system. That said, some control procedures are automated in corporate accounting systems. This makes it possible to ensure high efficiency and prompt decision-making in business systems (Leybert, E.B., 2020).

According to the economist N. Karimov, the internal control system can be interpreted as a system for "maintenance of control over improvement of certain actions by employees . . . " (Karimov, 2007). According to N. Karimov, we can see that internal control is defined as control over employees.

According to Professor L. Sotnikova, the primary objectives of internal control are to ensure reliability of information reflected in accounting records, and the integrity of assets and accounting records of a commercial entity (Sotnikova, 2015). Since the concepts of control and internal control are disparate, the efficient internal control system is an integral part of the management system of an enterprise.

According to the English scientist R. Dodge, "internal audit, which is conducted pursuant to the decision of the company management for the purposes of control and analysis of entrepreneurial activity, is an integral part of control internal" (Dodge, 1992). According to this economist, internal audit is shown as an integral part of internal control with a very narrow meaning.

Another concept of management methodology is based on the principle of its division into general and specific types, which does not narrow control down to certain research methods, but implies the use of dialectical principles and general scientific methods of cognition (Abdusalomova, 2017a; Abdusalomova, 2017b).

B. Khamdamov, "Internal control is a continuous day-to-day activity, which, in its turn, enables correct accounting" (Khamdamov, 2005).

According to some experts, the internal control system only has effect in cases where the managers deny that "internal economic control in the company is carried out by the internal audit service (special department)" and "control is appointed to the auditing committee based on public announcements" (Krupennikov, 2009). In our opinion, internal audit is only one of elements of internal control system.

In our opinion, internal control is an independent external activity on the provision of reliable advisory services for the benefit of an enterprise and efficiency improvement.

The analysis of the place and the role of internal control in the management of the iron and steel enterprises has made it possible to conclude that internal control is necessary at all stages and levels of management, primarily in the manufacture of finished products and industrial process management. However, the level of internal control carried out by the iron and steel enterprises may be insufficient due to a number of objective and subjective reasons.

Even if high-quality internal control has been established, in the process of manufacture of products, additional external evaluation of the status of internal control should generally be carried out; moreover, this should be done by independent well-qualified personnel. Such experts in the iron and steel enterprises should be part of special services exercising control functions, i.e. internal control service and internal audit service.

The mission of the internal control service in the management system of an iron and steel enterprise is to develop a reliable and functional internal control system, scientifically grounded methodology, regular assessment of the current state of internal control, its direct introduction in the absence of internal control in certain areas, preparation of internal report with the presentation of findings and recommendations on improvement. The author's research has been used as a basis for the recommendation of an algorithm of preparation of internal accounts in the iron and steel enterprises (Figure 7.1).

The objectives of preparation of internal reports based on internal control include: determination of the reliability of accounting (financial) statements, legality and expediency of their economic activity; rational organization of accounting and reporting; provision of internal users with timely information about results of control, which allows them to pass efficient economic decisions aimed at the rapid development of production and the increase in the level of profitability of economic responsibility centers; examination of accounting and financial statements with a view to determining their compliance with current regulations, legal status, documents of

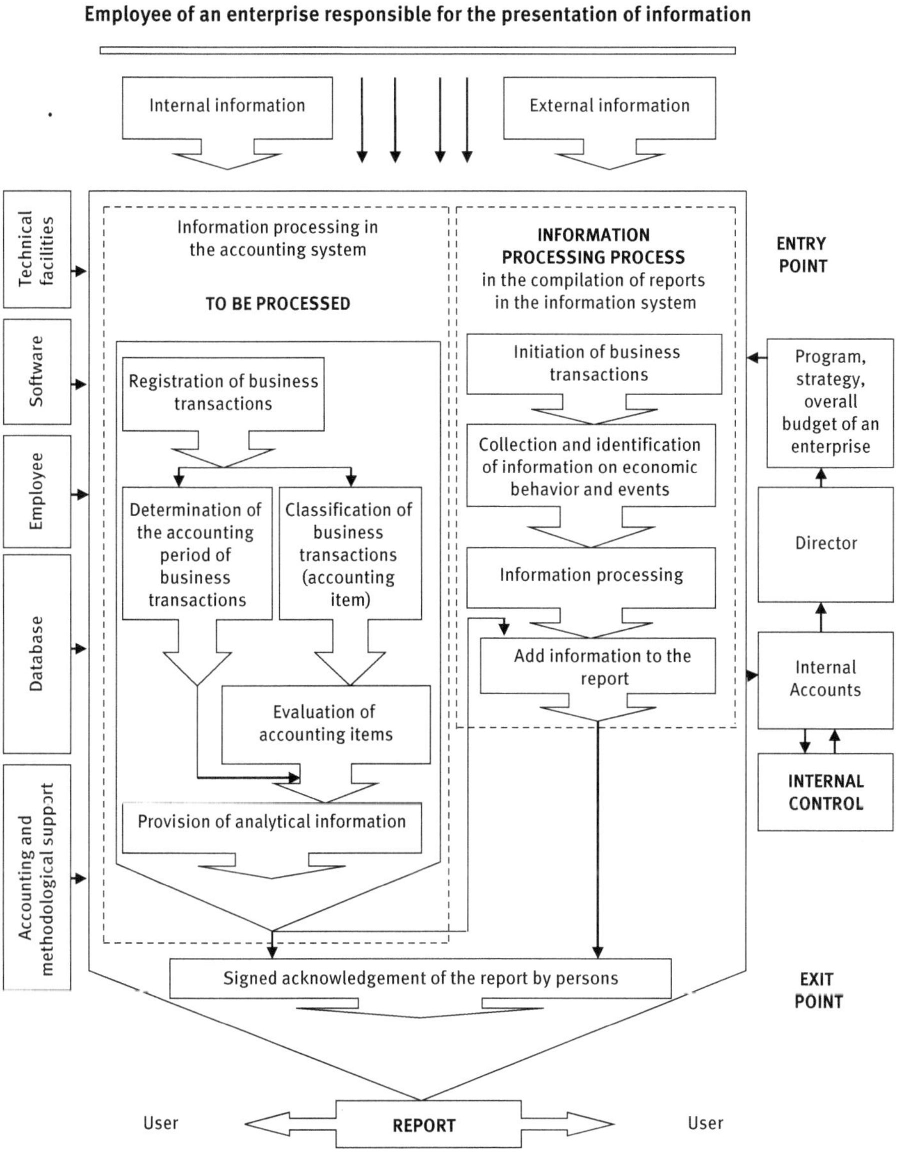

Figure 7.1: Algorithm for the generation of internal reports.
Source: developed and compiled by the author.

association, industry-specific and organizational characteristics of an enterprise and substantiation of proposals on removal of identified shortcomings; compliance audit of the established work schedule and management decisions by an enterprise; identification of resources to reduce the cost of production of goods, production of work, and provision of services.

Internal control in the iron and steel enterprises is managed by the management team, and in the implementation of business transactions all employees must exercise their functions in the most effective manner. Moreover, every employee must know who he or she is subordinate to and who he or she may place obligations on. Each employee must be fully aware what he or she is responsible for and what criteria will be used for assessment of results of his or her work.

Internal control over outgoing inventory in the iron and steel enterprises includes: timely completion of relevant documents for the manufacture of metal products of various grades; maintenance of control over its safety at the warehouses of an enterprise; timely coverage of activities on delivery and sales of finished products and customer accounts; ensuring internal control of production and sales plan. The automation of activity is necessary for improving accuracy of calculations and increasing efficiency of such activity as well as introducing the efficient internal control system.

Internal control of the management accounting system is carried out in a proper manner, including the timely identification of various management risks (for example, systematic errors and corrupt practices on the part of employees) and the development of measures to prevent them.

At the same time, it is necessary to extend the sphere of internal control based on the accounting and analytical subsystem, organizational structure of management, personnel management and other elements of organization. This approach necessitates full coverage of all elements of management methods of an iron and steel enterprise, since the failure of one element prevents from generating a broad description of an enterprise and impedes effective decision-making.

Currently, the methodology for some of objects of internal control in the management accounting system listed above, including personnel management methodology, is incomplete and prevents from organizing adequate internal control.

Conclusion

In conclusion it should be pointed out that efficient operation of an iron and steel enterprise is determined by the optimization of its organizational structures and segments, which, in turn, requires the improvement of the accounting and analytical subsystem, the internal control system of an enterprise.

In this regard, the improvement of internal control plays a special part in the management accounting system. Being a relatively new division of economic knowledge in the country, the management accounting system, the specificity of internal control and its versatility necessitate the use of unconventional means.

Improvement of the management accounting system through the implementation of goals and objectives of internal control enables the logical and critical analysis of management subsystems (sales, supplies, pricing, etc.) and systemic problem approach. The identification of characteristic features and development trends of each subsystem generally requires the identification of global development problems of an enterprise and visualization of the accurate "financial position" of an economic entity.

References

Abdusalomova, N.B. (2017a). Budgeting as a component of the management cost accounting system. In Necheukhina, N.S., Buyanova, T.I. (Ed.) Accounting, analysis and audit: the current state and prospects for further development: proceedings of the VIII International research-to-practice conference. Yekaterinburg, Publishing House of the Ural State University of Economics.

Abdusalomova, N.B. (2017b) The role of internal control in management accounting. In Necheukhina, N.S., Buyanova, T.I. (Ed.) Management of social and economic development of regions: problems and ways of solving them: collection of scientific articles of the 9th International research-to-practice conference. Yekaterinburg, Publishing House of the Ural State University of Economics.

Abdusalomova, N.B. (2019). Principles of ties of internal control and management accounting systems at the enterprises of black metallurgy. International Scientific Journal Theoretical & Applied Science, 2(1), pp. 385–391.

Astafeva, O.V., Astafyev, E.V., Khalikova, E.A., Leybert, T.B., Osipova, I.A. XBRL Reporting in the Conditions of Digital Business Transformation // Lecture Notes in Networks and Systems. – 2020. – No. 84. – P. 373–381.

Dodge, R. (1992). The Concise Guide to Auditing Standards and Guidelines: translated from English. Moscow, Finansy i Statistika; YUNITI Publishing House.

Karimov, N.F. (2007). Tizhorat banklarida ichki auditni tashkil kilish uni uslubiyotini takomillashtirish muammolari: a monograph. Tashkent, Publishing House of the Banking and Finance academy of the Republic of Uzbekistan.

Khamdamov, B.K. (2005). Audit iktisodi: a monograph. Tashkent: Nauka Publishing House.

Krupennikov, V.M. (2009). Improving the efficiency of internal control of a manufacturing enterprise amid the crisis: thesis for the degree of Candidate of economic sciences. Moscow, Nauka Publishing House.

Leybert, Tatiana B.; Khalikova, Elvira A. Current Tendencies of Transformation of the Russian Practice of Decision Making in Business Systems // Leading practice of decision making in modern business systems: innovative technologies and perspectives of optimizatio. – 2020. – P. 13–26.

Paramonova, L. (2012). Typology and patterns of organization of internal control of an economic entity. Resyrsy. Informatsiya. Snabzhenie. Konkurentsiya, 2(1), pp. 362–365.

Sheviakova, S.R. (2017). Internal audit as a tool for the improvement the operation of an enterprise. Molodoy Uchenyi, 11(145), pp. 295–297.

Sotnikova, L.V. (2015). Assessment of the current state of internal control: a practical guide. Moscow, YUNITI-DANA Publishing House.

Voronina, L.I. (2014). Audit: theory and practice: a textbook, 3rd edition, revised, Moscow, Omega-L Publishing House.

Ismatilla T. Ydyrysov, Keneshbek B. Yrysov, Emir Z. Tuibaev,
Zhenishbek A. Kochkonbaev and Oyatilla A. Umurzakov

8 Tactical and Technical Solutions for Intraoperative Critical Situations

Introduction

Intraoperative extreme situations are the one of the integral aspects of extreme surgery for severe injuries and polytrauma (Aghajanyan, 2006) (Bisenkov et al., 2002) (Mallaev, 2002). The ability to find a way out of it is the most difficult and important requirement for a surgical team (CS), which is forced to make life-threatening decisions with an acute shortage of time and necessary information (Ashimov, 2016) (Goncharov et al., 2003) (Mamanazarov et al., 2003).

It's well known that there are practically few scientific-methodological, educational-methodological materials that consider the factor as the autonomous and collegial resolution of ES from the victim directly on the operating table. Meanwhile, it's important to find answers to such questions. What are the basic principles of joint actions of the surgical team in such conditions? What are the most common approaches and models for understanding and correcting an extreme situation?

Undoubtedly, the surgical team in the process of providing operative and resuscitation care to victims in an emergency situation also accumulates the corresponding experience and skills (Davydov, 2006) (Kozhakmatova et al., 2002) (Osmonaliev et al, 2006). However, this doesn't allow formulating theoretical principles that could be applied in an extreme situation of risk, an extreme situation of extreme necessity and an extreme situation of a forced experiment with equal success.

The whole catch is that an extreme risk situation, an extreme situation of extreme necessity and an extreme situation of a forced experiment have specific features, special properties that the surgical team should distinguish before correcting it (Akinipin et al., 2003) (Ashimov, 2012). This is why a generalized approach to managing these emergencies, using skills and techniques common to a large number of cases, should become an important new component of the training of the surgical team.

Ismatilla T. Ydyrysov, Emir Z. Tuibaev, Oyatilla A. Umurzakov, Osh State University, Osh, Republic of Kyrgyzstan
Keneshbek B. Yrysov, Zhenishbek A. Kochkonbaev, Kyrgyz State Medical Academy I.K. Akhunbaeva, Bishkek, Republic of Kyrgyzstan

Materials and Methods

The material for the analysis was the relevant data on the provision of surgical and resuscitation care to 100 victims with severe injuries (n-16) and polytrauma (n-84) in an extreme situation.

As you can see from the Table 8.1, resuscitation with an operative component was performed in 97% of the victims. 3% of the victims died on the background of resuscitation measures without surgery. Moreover, 1 victim was in an extreme situation of extreme necessity, and 2 – in an extreme situation of a forced experiment. The condition was assessed as extremely serious, and the severity of injury or injury in this category of victims was assessed as incompatible with life.

Table 8.1: Distribution of victims with an extreme situation by outcome, depending on the performance of the operation.

ES	n	Including:			
		Resuscitation with an operational component		Resuscitation without an operative component	
		n	The number of died persons	n	The number of died persons
ESD	60	60	1	–	–
ESCN	24	23	7	1	1
ESVE	16	14	12	2	2
Total:	100	97	20	3	3

Results

Almost all victims in an extreme situation had a combined shock of varying degrees. As a rule, the shock of 28 victims of extreme risk situations was regarded mainly as grade I–II, and in 14 – grade III–IV shock.

Meanwhile, in the group of victims in an extreme situation of extreme necessity, in most cases, the shock of the III–IV degree appeared.

Moreover, 5 victims had grade III shock, and 9 – grade IV shock. In the group of victims in the extreme situation of the forced experiment, 5 of them had the shock of the III degree, and 11 victims had the shock of the IV degree.

Hemothorax of varying degrees was observed in 14 (14%) patients who were in the ESD, ESCN and ESVE. If in the group of victims with ESD hemothorax was observed only grade I, then in the group of victims with ESD and ESD – II–III grade.

Pneumothorax was observed in 2 victims with ESD, in 8 victims with ESD and in 8 victims with ESVE.

27 patients had hemoperitoneum of varying degrees. Moreover, 3 patients with ESD were diagnosed with grade I hemoperitoneum, 2 patients with ESD and 3 victims with ESVE also had such a degree of bleeding into the abdominal cavity, which was proven on the operating table. 2 victims were also with ESCN and 2 victims were with ESVE, hemoperitoneum was equal to grade II. Hemoperitoneum grade III was diagnosed in 6 victims with ESCN and 9 victims with ESVE.

The cause of the lethal outcome is, of course, decompensated shock, complicated by acute massive blood loss, ATSF and MOFS. As a result of autopsy, the cause of mortality in 11 patients (1 with ESD, 4 with ESCN and 6 with ESVE) was acute massive blood loss; while the reason of death of 8 victims was ACF (3 with ESD, 5 with ESVE). The cause of death 3 victims was ONE (1 – with ESCN, 2 – with ESVE).

The most frequent tactical omission of doctors was that the victim in the ES was initially hospitalized in the intensive care unit, and then he was transported to the operating room from there (41%).

It should be noted that having hospitalized a patient along route 2, i.e., from the emergency room to the intensive care unit, surgeons often «lose» sight of these victims. Resuscitators, while conducting syndromic therapy, don't focus on the threatening consequences of trauma, especially of an intracavitary nature. Thus, the timing of the required operation is postponed for some time, and at the same time, as you can know, the prognosis becomes doubtful or even unfavorable.

Another, no less frequent, tactical omission was that the time for the provision of resuscitation measures was unjustifiably prolonged to the detriment of the urgency of the operational component of this resuscitation, which is currently required (38%).

Insufficient intraoperative using of the diagnostic capabilities of chronic bronchitis occurred in 20% of cases, while violation of the established procedure for the using of diagnostic methods, prognosis and surgical resuscitation benefits from the side of chronic bronchitis occurred in every fourth victim (25%).

Insufficient and untimely using of the advice of the relevant specialists in the intraoperative diagnosis, prognosis and treatment of victims on the operating table is noted in 12% of cases.

Not only on the part of anesthesiologists-resuscitators, but also on the part of surgeons, such miscalculations were made as underestimation of the degree of homeostasis disturbance and also the nature of damage to internal organs (32%), and as a result, an unjustified expansion of the volume of surgical intervention to the detriment of the completeness and timing of resuscitation measures (26%).

Meanwhile, there were also opposite tendencies in decisions. In particular, in 36% of cases, there was an unjustified reduction in the volume of the operation at the insistence of anesthesiologists-resuscitators, or an unjustified acceleration of the operation time on the part of surgeons, contrary to the requirements of anesthesiologists-resuscitators.

In 31% of cases, an incorrect indication for simultaneous operations was fixed; there were disagreements between surgeons and anesthesiologists-resuscitators. In 32% of cases, the nature of intraoperative complications was underestimated in time on the part of CP members.

Serious speculation is the fact that in a certain number of victims, a complication factor is an underestimation on the part of CP of either the nature and degree of homeostasis disturbance, or the nature of the victim's traumatization, or underestimation of intraoperative complications. In most cases, there were some disagreements between surgeons and anesthesiologists-resuscitators.

All the text above testifies to the fact that the quality of these tactical decisions in a particular case is clearly insufficient. If we take into account the thesis that surgery is strong in its tactics in these, at first sight, inconspicuous factors that there is a serious reserve for improving the quality of surgical and resuscitation care for victims in ES.

We would like to note that almost every third or fourth victim has erroneous tactical decisions. A relatively low experience of intraoperative diagnostics took place in ¼ of the victims. The following inconsistent misconceptions of doctors are noted in the text above, as violations of the technology of intraoperative diagnosis and prediction.

It can be assumed that such tactical miscalculations are based on the self-confidence of CB members, the lack of coordinated activity, and moments of reassessment of their capabilities. This is confirmed by unjustified surgical activity or unreasonable delay in the timing of the operation, complication or simplification of the nature of the operation.

As you know, one of the important tactical moments in the provision of operational and resuscitation assistance is the implementation of the necessary RIC. The RIC has an important potential for resuscitation assistance to the victim during the operation.

So, in tactical terms, the provision of operational and resuscitation assistance to the victims in the ES was clearly insufficient.

Among the technical defects, special attention should be focused on the following groups of misconceptions: first, the wrong choice of the sequence of surgical intervention, which took place in almost every fourth victim with ES (25%) and also an inappropriate at that time surgical technique, which led to an increase in weight without that serious condition of the victims (22%).

Unfortunately, there were also such miscalculations as in a hurry, accidental damage to organs, tissues, blood vessels, which also contributes to the aggravation of the victim's condition (1%), as well as insufficiently reliable surgical hemostasis against the background of critical hypovolemia and post-hemorrhagic anemia in victims (13%).

We formulated the following summary judgment on the basis of the characteristics of the tactical and technical aspects of the decision of the ES in the provision of operational and resuscitation assistance to the victims in the ES.

Conclusion

1. It's necessary to focus on such an integral indicator as the degree of uncertainty, the degree of threat realization, the degree of predicting the consequences of choosing one or another alternative to surgical intervention while determining an adequate intraoperative assessment and resolving ES.
2. It's necessary to use tactics aimed at eliminating dangerous consequences, at stabilizing the crisis, focusing on such integral indicators of the functions of vital organs and systems as an indicator of changes in the cognitive status of the victim in ES.
3. The most frequent tactical omission of doctors was the fact that the victim in the ES was initially hospitalized in the intensive care unit, and then from there he was transported to the operating room (41%).
4. The time for the provision of resuscitation measures is unjustifiably prolonged to the detriment of the urgency of the currently required operational component of this resuscitation (38%).
5. Insufficient intraoperative using of the diagnostic capabilities of chronic bronchitis took place in 20% of cases, while every fourth victim has the violation of the established procedure for the using of diagnostic methods, prediction and surgical resuscitation (25%).
6. An incorrect indication for simultaneous operations was fixed in 31% of cases and there were disagreements between surgeons and anesthesiologists-resuscitators.

References

Aghajanyan, V.V. Organizational and tactical aspects of inter-hospital transportation of critically ill patients with polytrauma. Aghajanyan, A.B. Shatalin, S.A. Kravtsov // Polytrauma. – 2006. – No. 1. – pp. 18–27.

Akinipin, A.B. Resuscitation support of victims in emergency situations / A.B. Akinypin, Z.A. Albakova, Yu.A. Kotov. – Vseros. center honey. disasters "Protection". – M., 2003 .-- 49 p.

Ashimov, I.A. Riskology: dilemmas, judgments, decisions (III volume of selected works "BIOphilosophy") / IA Ashimov. – B., "Ilim", 2012. – 272 p.

Ashimov, I.A. System / I.A. Ashimov. – B., 2016 .-- 115 p.

Bisenkov, LN, et al. Emergency surgery of the chest and abdomen / Guide for doctors / LN Bisenkov, PN Zubarev, V.M. Trofimov. – M. – St. Petersburg: Hippocrates, 2002 -- 302 p.

Goncharov, S.F. Organization of medical support of the population in conditions of armed conflicts: guidelines / S.F. Goncharov, B.V. Bobiy, V.I. Kryuklov. – M .: VTsMK "Zashchita", 2003. – 78 p.

Davydov, V.M. Methodology for assessing the functioning of the optimization of the system of military medical education in modern conditions / V.M. Davydov: Author. dis. . . . doct. honey. sciences. – M., 2006 .-- 48 p.

Khorram-Manesh, A. Management of traumatic liver injuries without a valid trauma system / A. Khorram-Manesh, B. Pourseidi // Prehosp Disaster Med. – 2009. – No.24. – V.4. – pp. 349–355.

Kozhakmatova, G.S. On the issue of training and retraining of specialists in the provision of emergency medical care in peacetime disasters / G.S. Kozhakmatova, S.Sh. Toimatov, A.M. Mallaev, K.A. Shukurbaev // In the book: Actual problems of war surgery and disasters in mountainous conditions. – Bishkek, 2002. – pp. 33–34.

Mallaev, A.M. Clinical riskology / A.M. Mallaev: ed. I.A. Ashimova. – Bishkek, 2002 .-- 182 p.

Mamanazarov, DM, Organization of medical care and treatment of victims in extreme situations [Text] / DM Mamanazarov, NM Kurbanov, KH Khudaiberdiev. – Tashkent. Publishing house of medical literature. Abu Ali Ibn Sina, 2003 .-- 163 p.

Osmonaliev, D.M. Osmonaliev D.M., Abylgaziev I.T., Faizullaev R.A., Kichinegulov T.I. One-stage operations for combined injury // TsAMZh. 2006. – Vol. XII. – Number 3. – Appendix 3. – pp. 33–34.

Part III: **Sustainable Development in Industry 4.0**

Shakhlo T. Ergasheva and Rano A. Mannapova

9 Digital Modernization in Enterprises of Agricultural and Water Management Industry

Introduction

One of the major problems of national agricultural producers is improving their competitiveness. Successful management of the competitive position of an agricultural enterprise can be organized through the change of approaches to the organization of production and management. The basis of such approaches is the introduction of integrated digital systems based on computer complexes for the management of an enterprise. Such systems constitute the program of implementation of methodology based on management standards which are popular throughout the world. Moreover, a management standard shall be understood to mean the standard of functional consideration of processes (production, logistics, finance, marketing) and their results with reference to each other. These standards make it possible to harmonize and synchronize processes in real time.

As we know, digital economy is a paradigm that is based on digital technologies, related to e-business and e-commerce, as well as produced and marketed digital goods and services. Settlements for services and goods in the digital economy are often made by electronic money – digital currency. In this regard, the opportunities of the digital economy and blockchain in Uzbekistan are considered to be very promising. On September 2, 2018, free-market activity of companies in the area of crypto-asset turnover and blockchain technologies has been launched in Uzbekistan. Moreover, these technologies are being implemented in the public sector on the terms of public-private partnership. In addition, the Digital Trust foundation has been established, the objectives of which include attraction of investments, implementation of promising projects in the field of development of the digital economy, including those related to the introduction of blockchain technologies, namely: in the field of the automated system of registers for the State Centre of Expertise and Standardization of Drugs, healthcare products and medical equipment (President of the Republic of Uzbekistan, 2020).

Shakhlo T. Ergasheva, Tashkent State University of Economics, Tashkent, Republic of Uzbekistan
Rano A. Mannapova, Tashkent Institute of Finance, Tashkent, Republic of Uzbekistan

https://doi.org/10.1515/9783110654486-009

Methodology

The need for the introduction of digital technologies based on current management standards is obvious. However, most enterprises pay inadequate attention to this issue. Those same enterprises that introduce in-house automated management systems, and the number of such enterprises is constantly increasing, prove that there is the need to develop digital technologies in other enterprises by their successful economic activity.

Scientifically grounded recommendations on the optimization of distribution and the use of resources were given for the first time ever in 1939 by Kantorovich, L.V. (Mathematical methods of organization of production planning. Publishing House of the Leningrad State University. -1939.). At a later stage, many scientists greatly contributed to the development of automated management systems. Their works enabled successful use of automation for the management of enterprises. Given the transition to market relations, former information systems require serious development; at the same time, digital technologies are already way ahead throughout the world. Therefore, the works of many scientists are mainly aimed at using foreign approved management standards in digital systems. In fact, recommendations of many scientists are focused on the problems of implementation and operation of digital management systems. In turn, the issues of assessment of their economic efficiency faded into insignificance, despite the fact that conventional methods of assessment of economic efficiency of digital management systems are characterized by severe restrictions. Hence, scientific research activities dealing with the resolution of this complex problem, are highly relevant, as noted by Vanchukhina et al. (2018, 2020), Matveev (2012), Chumakov (2019), Khalikova and Stanishevskiy (2020).

Today, digital modernization is more developed in advanced industries, including oil and gas industry, since it is the driver of international economies. The process of digitalization in various industries has changed the technology of implementation of business processes, thereby improving economic efficiency of enterprises. The capability of rapid processing of Big Data in real time becomes the basis for the assessment of efficiency of enterprises (Peskova, D. R., 2019).

Recently, there has been a tendency towards the unification of various types of digital technologies into a single integrated set. Special part in it is played by methodologies, reflecting the management functions implemented in software of information systems. Such systems may be referred to as digital management systems (DMS). DMS are designed as a complex information technology and program complex. It supports a single method of data submission, satisfies informational and computational requirements of experts. Theoretical analysis has made it possible to clarify the definition of DMS. Digital Management System (DMS) of an enterprise is an information system based on the computer complex, integrated into the business processes of an enterprise and managing a certain resource set (production capacity, financial assets, personnel, material supply, etc.) of an enterprise. Therefore, DMS represents the system for the transformation of information into a production resource.

Results

The analysis of levels of development of digital technologies in the country shows that production volume in basic industries that use software products is low compared to other developed countries. Accordingly, indigenous companies have less funds which they can invest in the purchase of software products. In addition, in the economy, the demand for computer programs contributing to the increase in performance, is limited by the fact that the increase in productivity in many industries does not lead to the increase in the level of profitability or market success. In actual fact, the number of high-producing enterprises in the economy is generally small. As a result, companies invest disproportionally small share of their revenue in the purchase of software products and design services.

In order to analyze the current state of development of digital systems in our country, the activity of many enterprises of various forms of ownership has been studied, differing in the scale and the range of activity in various sectors for the recent five years. The analysis and assessment of their position have made it possible to identify trends and patterns of development of information management systems in enterprises, which are based on the status of sectors. Table 9.1 presents brief description of the current status of DMS of enterprises by sectors.

Table 9.1: Brief description of development of digital technologies in sectors of economy.

Sector	Level of development of DMS	Current status of the pool of computer facilities
Aircraft engineering industry	DMS are used by enterprises	Additional implementation of modern computer facilities is required
Motor industry	Modern DMS are implemented in enterprises	Modern computer facilities are used by most enterprises
Consumer goods industry	Uneven level of development of DMS	Many enterprises require significant additional implementation of modern computer facilities
Medical industry	Uneven development of DMS	Insufficient number of computer facilities in enterprises

Table 9.1 (continued)

Sector	Level of development of DMS	Current status of the pool of computer facilities
Oil extracting industry	Enterprises have low amount of computer workstations, corporate DMS should be introduced	Enterprises use modern computer facilities
Food processing industry	Enterprises have low amount of computer workstations, MRP II class systems should be introduced	Some enterprises have sufficient number of modern PCs
Trade industry	Enterprises have standard amount of computer workstations, some of them use marketing systems and Internet tools	Many enterprises require additional implementation of computer facilities and barcode scanners
Chemical industry	Modern DMS are used by enterprises	Enterprises are poorly equipped with modern computer facilities
Power industry	Modern DMS are used by enterprises	The number of computer facilities in enterprises is insufficient; they should be upgraded
Agricultural and water management industry	DMS are not used	Implementation of computer facilities is required

Source: developed and compiled by the authors.

In the current economic environment, the level of development of DMS of an enterprise depends on the overall status of the sector in which it operates. If the industry is in critical condition, the enterprises of this industry usually fall short of funds for the development of modern DMS. Therefore, the core of the concept of development of DMS must be relying on the industry-specific approach, and on the optimality from the "price – quality" standpoint. It means that it is necessary to evaluate the financial capacity of enterprises for the development of DMS and the necessary level of coverage of resources of an enterprise by the system to integrated management to the greatest possible extent. Enterprises with various level of provision of tangible and financial resources operate in various industries in different ways. They have different ranges of activity and different manufacturing cycles; therefore, the levels of development of information systems are different as well. The quantitative assessment of the nature and the form of the impact of financial and economic indicators of enterprises and the level of development of information management systems taking into account their industry-specific features can be determined using the following equation:

$$NA = A0 + A1 * OS + A2 * TP + A3 * B \qquad (1)$$

where NA is intangible assets (thousand soms); A_0 is the absolute term of equation, determining the influence of unaccounted factors; A_1, A_2, A_3 are reduction factors; OS are capital assets (thousand soms); TP is commodity output (thousand soms); B are assets of an enterprise.

If we insert relevant financial indicators of a particular enterprise into the equation of the respective group, we can make findings of the financial capacity of introduction of DMS.

Regression equation can be used to determine the economic efficiency of automated management systems in enterprises, but the main emphasis should be placed not on intangible assets, but on the indicators of availability of digital technologies.

$$NU = A0 + A1 * NA + A2 * OS + A3 * B \qquad (2)$$

where NU is the availability of DMS.

In the calculation of the standard efficiency factor (K) of the information system, it is necessary to determine the mean period between upgrades in years for personal computers and software, intervals between training of personnel involved in the information system, as well as the price for these components of the information system (Table 9.2).

Table 9.2: Upgrades of components of DMS.

Cost items	Mean period between upgrades, years	Mean price, thousand soms	Share of the total cost value, %
Personal computer	2.25	829,902	59.3
Software	2.86	314,428	22.5
Training of one person	4.00	188,400	13.5
Miscellaneous costs	0.08	66,645	4.7
Total		1,399,575	100

Source: developed and compiled by the authors.

Based on percentage ratio of costs and mean periods between upgrades of components of the information system, mean payoff period (C = 2,522) and the efficiency factor of the information system (K = 0.397) can be determined.

Conclusion

Therefore, the identified development trends and patterns of digital management systems in enterprises make it possible to single out enterprises with high-tech production capacity that contributes to the rapid practical approval and further improvement of modern DMS. Distinctive industry-specific features of enterprises and their relationship to the level of development of digital management systems primarily consist in technological capabilities of the majority of enterprises in this sector and the overall financial and economic performance of the sector.

Therefore, it is necessary to develop new approaches to the estimate of necessary costs for the use of information management systems of enterprises based on comparison of economic efficiency indicators. Subsequently, it is necessary to justify standard efficiency factors of costs on the development of informative systems of enterprises, taking into account both the structure of costs on information systems and the differentiation of payoff periods of individual components of information systems. All of the above may contribute to the expeditious introduction of DMS and to the multiplication of existing best practices in the use of digital management systems by various water management and agricultural enterprises. This approach can be used as the basis for the development of recommendations to improve the financial position and the projected amount of efficient investment with the use of digital technologies in enterprises of agricultural and water management industry, taking account into industry-specific features.

References

Chumakov, V. (2019). Digital leadership strategies. V Mire Nauki, 10(1),70–77.
Khalikova, E.A., Stanishevskiy, E.E. (2020). Current trends in development of the digital economy in the context of innovative development of the problem and development trends in the innovative economy: international experience and Russian practices; proceedings of the VIII International research-to-practice conference of March 30, 2020, pp. 253–257. Ufa, Publishing House of the Ufa State Petroleum Technological University.
Matveev, I.A. (2012). Electronic economy: essence and stages of development. Upravlenie Ekonimicheskimi Sistemami: electronic academic periodical, 6 (42), 48–59.
Peskova, D. R.; Khodkovskaya, J. V.; Charikov, V. S.; Sharafutdinov, R. B. Development of business environment of oil and gas companies in digital economy // EUROPEAN PROCEEDINGS OF SOCIAL AND BEHAVIOURAL SCIENCES. – 2019. – No.57. – P. 1205–1212.
President of the Republic of Uzbekistan (2020). Resolutions of the President of the Republic of Uzbekistan No. 3832 of July 3, 2018 "On measures aimed to develop the digital economy in the Republic of Uzbekistan". URL: https://lex.uz/docs/3806048 (accessed: 03.09.2020).

Vanchukhina, L.I., Leybert, T.B., Khalikova, E.A., Luneva, N.N. Modern approaches to operational planning in oil refinery using the PIMS software product // Quality – Access to Success. – 2018. – No.19 (S2). – P. 123–129.

Vanchukhina, L.I., Leybert, T.B., Halikova, E.A., Rudneva, Yu.R. (2018). Development of Application of the Method of the Analysis of Scenarios in Profitable Approach to Estimation of Cost of the Oil Company. Bulletin Social-Economic and Humanitarian Research, 1 (1),42–54.

Rasul O. Kholbekov and Feruzakhon R. Kholbekova

10 Role and Value of Production Accounting in Providing the Company's Microeconomic Stability

Introduction

The level of microeconomic stability depends on the financial result of the activities of each enterprise. The financial result can be positive and negative. When analyzing the level of microeconomic stability of the enterprise, the main information base is the report "On financial results". In the preparation of the financial results report, production accounting records are used, and production accounting determines the production cost of products (works and services). Therefore, we believe that production accounting is essential in ensuring microeconomic stability of an enterprise.

Financial results are the final result of the enterprise. The financial results of the enterprise are characterized by indicators such as gross profit, profit from core activities, profit from financial activities, extraordinary profits and net profit. The profit of the enterprise is the main source of financing production and economic development of the enterprise. The financial result of the economic activity of an organization is determined by the profit or loss indicator formed during the calendar (economic) year. The financial result represents the difference between the amount of income and expenses of the organization. The excess of income over expenses means an increase in property – profit, and expenses over income – a decrease in property – loss. The financial result obtained by the organization for the reporting year in the form of profit or loss, respectively, leads to an increase or decrease in its equity.

The organization's activities include a set of facts of economic life related to the expenditure of material, labor and financial resources. The economic essence of such expenses is to reduce the economic benefit of the organization, which is the result of the disposal of assets (cash and other property) and the emergence of obligations that lead to a decrease in the capital of this organization, with the exception of reducing deposits by decision of the participants (property owners).

Rasul O. Kholbekov, Feruzakhon R. Kholbekova, Tashkent State University of Economics, Tashkent, Republic of Uzbekistan

https://doi.org/10.1515/9783110654486-010

Materials and Methods

After the introduction into practice of the Republic of Uzbekistan, the Regulation "On the composition of the costs of production and sale of products (work, services) and the procedure for generating financial results" approved by Resolution No. 54 of the Cabinet of Ministers of the Republic of Uzbekistan dated February 5, 1999, the methods and methodologies for calculating cost have changed. According to the Regulation, expenses are divided into costs included in the production cost of products (works, services) and expenses of the period not included in the cost of products (works, services) and closed at the expense of profit (loss) of an economic entity.

The first group includes expenses that directly related to the production of products (works, services), due to the technology and organization of production. These include: direct and indirect material costs, direct and indirect labor costs, other direct and indirect costs, including production overheads. Costs are recognized in accounting under the following conditions:
- The expense is carried out in accordance with a specific contract, the requirements of legislative and regulatory acts, business customs.
- There is confidence that as a result of a specific operation, the economic benefits of the organization will decrease. Confidence that, as a result of a particular operation, the organization will reduce its economic benefits, is there if the organization transferred the asset or there is no uncertainty regarding the transfer of the asset.

With the introduction of foreign investment in the economy of the Republic of Uzbekistan, the integration of accounting accounts also began. New terms, principles, methods, accounts, and report types are displayed in the current account system. Now, accounting was introduced into the stage of direct participation not only in the economic entity, but also in monitoring, recording, analyzing and collecting information on the processes taking place in the business entity. As a result of the introduction of new terms, principles and methods, such accounts as accounting, financial, managerial and production accounts appeared in the system of current accounts.

In the period after the Second World War, the methods used in the United States of America did not meet the requirements of management because of progress in achieving scientific and technological achievements, aggravation of the socio-economic situation, internal and external competition, intensification of inflation and changing production methods. Thus, in the 1950s, new sections of the US gateway account appeared in the US account system.

Despite the fact that more than a century the world of finance and management accounts seems to be the same, there are different opinions about the difference and unity of this world. Until now, many scientists – Ibragimov and Karimov (1999), Karimov (2004), Khasanov (2003), Kholbekov and Necheukhina (2019), Lisovich and Tkachenko (2000), Satubaldin (1980), Sheremet (1999), Sokolov (1996),

Xolbekov (2005) – have formulated their recommendations on the criteria for distinguishing financial calculations, production and management accounts in their scientific works.

Results

Using the criterion of differentiation of financial and managerial calculations from the scientists above, we dwelled in detail on the differences between the financial and managerial accounts of "Production Accounting" and presented it in Table 10.1.

As can be seen from the table, there are general and unambiguous aspects of the organization of financial, managerial and production accounting.

Table 10.1: Definition and criteria of differentiation of financial, managerial and production accounts.

No.	Distinctive character and criteria	Financial accounting	Management accounting	Production accounting
1	Users of information	International organizations, financial institutions, investors, shareholders, government bodies and internal accountable developers	All trainees and staff at the level of internal management	Production, financial reporting and management decisions
2	Freedom of choice	It is limited by generally accepted accounting principles, rules, standards and guidelines	There are no restrictions on the amount of income received from decisions on good governance	Policies, rules and regulations of generally accepted accounting standards
3	Application of methods	Documentation, Inventory, Accounting, Duplication, Valuation and Calculation, Balance, Reporting	All methods of accounting and index, mathematical methods and analysis	Documenta- tion, inventtory, accountting, duplication, evaluation, repor-ting and partial costing
4	The unit of measure used	Monetary unit of measurement	Monetary, natural, time unit of measurement	Monetary and natural unit of measure
5	Organization	Required	Is voluntary, with the consent of the manager	Depending on the nature of the enterprise, with the consent of the manager

Table 10.1 (continued)

No.	Distinctive character and criteria	Financial accounting	Management accounting	Production accounting
6	Information accumulation period	After business operations	Before business operations	After business operations
7	Level of information reliability	Based on these documents	Based on reliable, predictable information	Based on real documents
8	Sources of report creation	Accounting records, primary documents, accounting registers	Accounting records, oral and written information, as well as testimony of the administration and decisions of the councils	Report of accountable persons and records of accounting, primary documents, accounting registers
9	Reporting period	In a certain period (annually or quarterly)	In necessary cases (time is not defined)	At the end of each production period
10	Level of transparency of information	Open to all. Can be published in the media	It is used only within the business entity. Not specified in external reports	To be kept secret
11	Cost accounting method	By object of expenditure	For costing articles	Constant and variable, by direct and indirect costs
12	Object of the account of expenses	The cost price within the business entity	Responsibility Centers	By location of costs
13	Participants in the report creation	Chief accountant or his deputy	Technologists, department managers, chief accountant and other responsible persons	Managers of the shop or depar-tment of produc-tion and an employee of the accounting department

Source: developed and compiled by the authors.

Discussion

Signs and criteria for this difference were reflected in the brochures and scientific studies of many scientists who expressed their views on this issue. We have included several proposals. Until now, this difference existed only between financial and managerial accounting, but we propose that industrial accounting be perceived as part of accounting and added to the signs and criteria of difference.

Scientists also expressed their opinion on the relationship between financial, managerial, production and tax accounting, which are types of accounting. Figure 10.1 shows how A.D. Sheremet describes the relationship of these elements in his book "Managerial Accounting" .

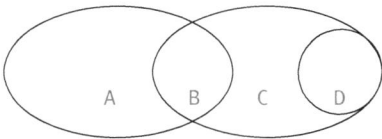

Figure 10.1: The relationship between types of accounting.
Source: developed and compiled by the authors.
Note: A – production accounting
B – financial accounting (for internal management)
C – narrow financial accounting (for external use)
D – financial accounting

It follows that A.D. Sheremet describes the existence of the above-mentioned accounting units and their relationship, which means that the management account is the link between financial and production records, and tax accounting is on financial accounting.

Figure 10.2 shows how G.M. Lisovich and N.Yu. Tkachenko presented the relationship in their book "Accounting Management Accounting in Agriculture and Processing Enterprises of the Agroindustrial Complex".

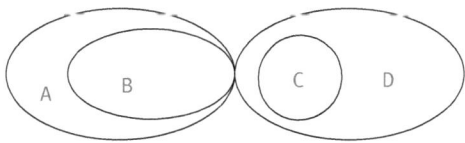

Figure 10.2: Relationship between types of accounting.
Source: developed and compiled by the authors.
Note: A – management accounting
B – production accounting
C – financial accounting
D – tax accounting

This figure shows that management accounting is closely related to production accounting, and that tax accounting is in financial accounting and that these accounts are to a certain extent interrelated.

In his book "Managerial Accounting: myth or reality", Y.V. Sokolov distinguishes between two types of accounting. Figure 10.3 shows their relationship.

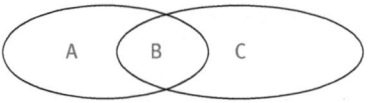

Figure 10.3: Relationship between accounting components.
Source: developed and compiled by the authors.
Note: A – is management accounting
B – accounting policy
C – financial accounting

As can be seen from this picture, the author has two components that make up the financial and management accounting, as the accounting policy of their interconnection.

Karimov, F. Islamov, A. Avlokulov in his book "Accounting" divide accounting into managerial, financial and tax accounting.

S.S. Satubaldin notes that production and financial reports have been merged, as all internal accounting processes are reflected in the main book.

In the opinion of A.K. Ibrahimov, "accounting of production costs can be carried out both in production and in financial accounting. If the production accounting is effective for the management of an economic entity and the determination of the production cost, the financial statements will be used to manage financial and commercial activities and prepare financial statements".

Many scientists also expressed opinions on "Production accounting". For example, C.T. Horgner and J. Foster. According to Foster's ideas, the key words are the terms "management accounting" and "production accounting". K. Drury shows the difference between "Managerial Accounting" and "Accounting for Production". A.D. Sheremet considers production accounting as a separate type of account, Y.V. Sokolov and other scientists pointed out that the term "Accounting" replaced the term "Production Accounting", which refers to the term "Accounting of production".

Some scholars (Astafeva, O.V. et al., 2020) think that all types of accounting – production, regulatory (financial and tax), and managerial – are integrated in the unified corporate accounting system. This is proved by transformation of separate accounting systems through quick development of the accounting and analytical systems based on creation of the unified information space, as well as digitalization of the process of formation and visualization of corporate accounting in the XBRL form in real time.

K. Djumaniyazov and B. Makhsudov in the monograph "Fundamentals of Management Accounting" argued that the grouping of accounting, the distribution of current and future expenses and the preparation of reports account for production accounting. A.A. Dodonov in his book "The Normative Textbook: What it should be" notes that normative calculations play a key role in organizing the accounting of production. In his view, the main task of production accounting is to control the movement and storage of semi-finished products. In the book of T.P. Karpova "Managerial Accounting" indicates that instead of the usual production accounts, the concept "Production costs" and "Product price calculation" arose. G.M. Lisovich and N.Yu Tkachenko considers production part of the financial statements. B.A. Khasanov also in his works adds an account to a separate type of accounting. In our opinion, the term "production accounting" is not a novelty, since this is the opinion of the above-mentioned scientists. Thus, the term "production accounting" is used in the accounting system and can be used again. It is only necessary to justify its methodology, methods and principles in practice and in scientific terms. Production accounting, being a part of the accounting system that summarizes information on production costs, determines financial results, manages decisions and monitors their effectiveness, and also serves as a source of information for financial and management accounting.

The cost accounting system serves as the estimated cost for each type of product and provides cost management, cost accounting, and production costs. In addition to the above distinctive features and criteria, the management account is more subjective.

The organization of his work depends on the wishes of the manager, and the requirements of the administration and expectations are also taken into account when making managerial decisions. Financial and industrial accounting are objective. Because they are reflected in the same way as events. Along with this, many scientists of the Republic of Uzbekistan, as the main source of managerial accounting, use the Resolution of the Ministry of Finance of the Republic of Uzbekistan No. 9 of January 27, 1995 amended by Resolution of the Cabinet of Ministers No. 54 of February 5, 1999 "Regulations on the composition of costs and financial results of production." In our opinion, this system is a source of financial and management accounting, that is, mandatory criteria. However, taking into account the characteristics of the network or business entities on the basis of this generally accepted provision, the guidelines set out in the accounting policy and is the source. In addition, some scholars refuse to cover the topic "Cost Accounting" in their textbooks.

Accounting for costs should be accounted for in the accounting, financial, managerial and production accounts. Only a theoretical point of view should be covered theoretically, i.e., the concept of cost, classification, accountability, write-offs, reporting and account registers, as well as in management and production accounts, in particular, costs are grouped, recorded and written off.

As can be seen from the above analysis and opinions, there are some problems in the development of national accounting and the creation of a unified accounting system in Uzbekistan. It is especially important to develop and implement theoretical, practical, scientifically sound proposals and recommendations for the development, creation and improvement of methodology and methodologies and principles of financial, managerial and production accounting, which are sections and branches of accounting. There are different views on the structure of accounting, which allows for more in-depth research and research.

Conclusion

Proceeding from historical sources, we would like to give our proposals to economists, using their ideas in this field. Due to changes in the development of science and technology, ownership patterns and state accounting policies, it is possible to change the organizational structure, methodology and accounting principles. But, as a system, it retains its role and its place in the economy of society. At the same time, we would like to remind you of the inadequate views of our country. For example, if some of our scientists continue to use the word "Financial Accounting" instead of "Accounting", accounting can also be part of the financial system. There are also views and concepts that the "financial account" is part of the finance.

Based on the research, we describe the organizational structure of accounting in Figure 10.4.

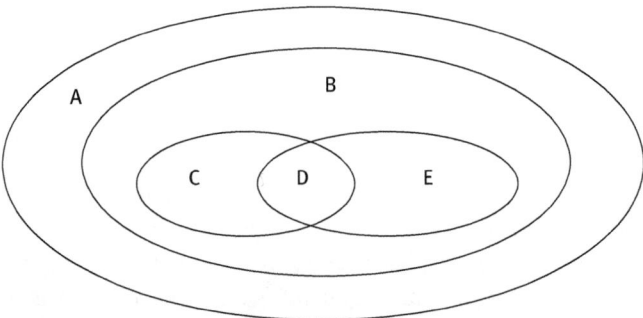

Figure 10.4: Organizational structure of accounting.
Source: developed and compiled by the authors.
Note: A – is the accounting system
B – Accounting policy
C – financial account
D – production account
E – management accounting

As you can see from this image, the production accounting is part of the accounting policy, as well as other types of accounting. Its organization should be within the accounting policy of doing business.

The production accounting data are displayed as a link between financial and management accounts.

It is also important to note that creating an account is not a mandatory requirement for all businesses. The criterion of difference is that it is based on the specific production characteristics of enterprises.

References

Astafeva, O.V., Astafyev, E.V., Khalikova, E.A., Leybert, T.B., Osipova, I.A. XBRL Reporting in the Conditions of Digital Business Transformation // Lecture Notes in Networks and Systems. – 2020. – No.84. – P. 373–381
Ibragimov, A.K., Karimov, A.A. (1999). Accounting of foreign investments. Tashkent, Uzbekistan.
Karimov, A.A. (2004). Accounting. Tashkent, Sharq.
Khasanov, B.A. (2003). Management accounting: theory and methodology. Tashkent, Finance.
Kholbekov, R.O., Necheukhina N.S. (2019). Accounting: Theory and Practice of Russia and Uzbekistan. Textbook. Yekaterinburg, Publishing House Ural state econ. University.
Lisovich, G.M., Tkachenko, N.Yu. (2000). Accounting management accounting in agriculture and processing enterprises of the agro-industrial complex, Rostov, Publishing Center "Mart".
Satubaldin, S.S. (1980). Accounting and cost of production in US industry. Moscow, Finance.
Sheremet, A.D. (1999). Management accounting. Moscow, FBK-PRESS.
Sokolov, Y.V. (1996). Accounting: from the beginnings to the present day. Moscow, Audit UNITY.
Xolbekov, R.O. (2005). Principles and methodology of organization of production accounting. Monograph. Tashkent, FAN.

Valery V. Gusev, Gamzat U. Magomedbekov, Gulnaz F. Galieva, Marina A. Gundorova and Zhanna A. Shadrina

11 Sharing Economy: How Digital Technologies Have Changed Economic Reality

Introduction

Socio-economic relations based on the sharing of resources and benefits have existed for a long time. However, the economic crisis of 2008 and the intensive development of digital technologies created a condition for the emergence of the sharing economy as a new form of economic activity of subjects.

Initially, only households that sought opportunities for a more economical, rational and efficient distribution of material resources through social networks were subjects of the sharing economy. However, the success of this interaction has pushed business structures to use the growing popularity of the trend. Firms have also begun to provide platforms for sharing resources and benefits, and digital technologies have greatly expanded the possibilities for sharing or using them. Indeed, the sharing economy has attracted more than $ 23 billion in venture capital investments over the past ten years and has shown strong growth (in 2014, the global sharing market was estimated at $ 15 billion, and by 2025 it will be close to $ 335 billion) (RBC, 2019).

The study of the conceptual prerequisites for the emergence of the sharing economy, as well as the identification of its specifics as a new form of socio-economic relations, is necessary to understand the essence of the changes taking place and predict the possible risks of a new formation for the economic system.

Methodology

Key research questions:
- What are the prerequisites for the emergence and development of the sharing economy?

Valery V. Gusev, K.G. Razumovsky Moscow State University of technologies and management (the First Cossack University), Moscow, Russia
Gamzat U. Magomedbekov, Dagestan State University, Makhachkala, Russia
Gulnaz F. Galieva, Ufa State Petroleum Technical University, Ufa, Russia
Marina A. Gundorova, Vladimir State University named after Alexander and Nikolai Stoletovs, Vladimir, Russia
Zhanna A. Shadrina, Kuban State Technological University, Krasnodar, Russia

https://doi.org/10.1515/9783110654486-011

– What are the specifics of sharing as a special type of socio-economic relations? What is the place of digital technologies in them?
– What are the economic benefits and risks of developing a sharing economy?

The basic concepts of the research are: 1) an industrial society is a society that has an economy based on a developed industry, as well as its corresponding social relations and political structure (Drucker, 1969; Bell, 1986); 2) an information society is a type of society where knowledge generation, information processing, and symbolic communication technologies are the source of productivity (Bodrunov, 2016; Tofller, 1999; Castels, 2000); 3) sharing is an activity involving the use of additional resources available to households to extract some social and material benefits through the use of digital technologies (Botsman and Rogers, 2010; Dillahunt and Malone, 2015; Heinrichs, 2013).

The works of Fromm (2000), Tawney (1920), Ekhart (1991), Porter (2005), Kotler (2012), K. Prahalad and Ramaswami (2006), Roshchina (2005), Auzan (2018), Pazaitis et al. (2017), Oh et al. (2013) were studied in the process of analyzing the prerequisites for the emergence of the sharing economy. The results of research by Botsman and Rogers (2010), Bauwen (2005), Codagnone et al. (2016), Petropoulos (2016), Wallsten (2015), Le Vine et al. (2014), Huber (2017), Hafermalz et al. (2016) were used to establish the role of digital technologies in socio-economic relations in the sharing economy. Identifying the risks of developing socio-economic relations in the sharing economy required studying the information provided in publications of Cohen (2015) and Dostmohammad and Long (2015).

Methods used in the research: method of theoretical analysis and synthesis, method of systematization, logical, dialectical methods, system approach. The information base of the research is the materials and reports of the World Bank (2018), Analytical center under the Government of the Russian Federation (2019), and of consulting companies PwC (2018), McKinsey (2016), Timbro (2018), as well as materials of scientific publications and international conferences.

Results

I Prerequisites for the Emergence and Development of the Sharing Economy

The evolutionary transition from an industrial society to an information society has caused a change in the theoretical concepts that serve as the basis of the industrial economy, and has also created prerequisites for the emergence of new forms of economic activity, including the sharing economy (Table 11.1).

Table 11.1: Prerequisites for the emergence and development of the sharing economy.

Conceptual prerequisite	Industrial society	Information society
Concept of attitude to property	The concept of possession	The concept of use
Consumer value creation model	A model for creating consumer value for a firm without its participation (Porter, 2005; Kotler, 2012)	Model of joint creation of consumer value (Prahalad and Ramaswami, 2006)
The paradigm of commodity relations	Concept of commodity exchange (purchase and sale)	The concept of a digital exchange (Dillahunt and Malone, 2015)
The transaction cost model	The usual model of economic institutions	Modified model of economic institutions

Source: compiled by the authors.

We will reveal in more detail the content of each conceptual prerequisite.

1 Transition from the Concept of Possession to the Concept of Use

Let's highlight three postulates of political economy that determine the transition from the concept of possession to the concept of use:
- Changing the nature of consumption. In an industrial society, consumption is the satisfaction of the vital needs of all members of society, and the purpose of production is to ensure the process of consumption and create added value (Roshchina, 2005). In the information society, consumption is a condition for satisfying a person's natural need for self-actualization (although it is inherent), and the purpose of production is to create conditions for human development, continuously increasing the level of self-actualization. Thus, the nature of consumption changes: increasing opportunities to meet the needs for material goods are replaced by a person's desire to meet spiritual and social needs, which determines shifts in the perception of property.
- Change in the perception of property. The nature of "possession" follows from the nature of private property: here the meaning is to acquire property and exercise the unlimited right of man to preserve what he has acquired. At the same time, the principle of possession excludes all others, it does not require a person to make any effort to preserve property or use it productively (Fromm, 2000). Overcoming the dominance of private property in the transition from an industrial society to an information society occurs not as a result of socialization of

production, but through the formation and formation of the institution of personal property. This is also due to changes in the functions of private property. Previously, private property provided an opportunity for a person to act as an economic subject, and also determined the conditions for disposing of the created product and profit. Thus, Tawney (1920) in the book "The Acquisition Society" notes that "to acquire, own and profit – the inalienable rights of the individual in an industrial society". As social development progresses, private property gradually dissolves, its boundaries are blurred, and personal property increases, and private property is now associated with the right to dispose of the created product and profit.

- Changing the relationship between property and freedom. Eckhart studied the relationship between property and freedom. The freedom of a person is limited to the extent that he is tied to property, to work, and to his "self" (Ekhart, 1991). Man is placed in his own path, his activities are unsuccessful, he does not realize his full potential, being tied to his own "I". In possession, the General attitude is important, not the various objects of possession (things, real estate, ritual, knowledge and thoughts): in the case of attachment to the objects of possession, they begin to restrict the freedom of a person and prevent his self-expression. In the context of the information society, freedom and self-expression of a person become the highest values. Thus, the concept of possession, which is characteristic of industrial society, becomes obsolete in the conditions of the information society.

2 Changing the Model of Creating Consumer Value

In the process of social development, there is a transition from the model of creating consumer value by a firm for the consumer without his participation (Porter, 2005; Kotler, 2012) to the model of joint creation of consumer value (Prahalad and Ramaswami, 2006). In their book "The Future of competition", Prahalad and Ramaswami analyze the structural changes caused by the convergence of industries and technologies, ubiquitous connectivity and globalization, and, as a result, the evolving role of the consumer from a passive recipient to an active co-creator of value. In the information society, firms can no longer create value on their own, and consumers interact with a network of firms and consumer communities to create value together. A personalized collaborative experience becomes a source of unique value for both consumers and firms.

3 Changing the Paradigm of Commodity Relations

In an industrial society, the exchange of products is carried out mainly spontaneously and in the form of commodity exchange (purchase and sale). In the information society, there is a gradual weakening of the role of cost characteristics that underlie exchange, and the continuous strengthening of value (various kinds of status, individual, subjective) characteristics preserves the element of spontaneity in the exchange of products and goods (Roshchina, 2005). Goods are exchanged on a global scale through digital platforms (Dillahunt and Malone, 2015).

4 Changing the Transaction Cost Model

In industrial society, the usual model of economic institutions operates: hierarchies, inter-firm networks, and the market. In the transition to the information society, the usual model is modified, and the elements of the modified model become: hierarchies with ultra-low control costs; company-free networks; markets with aggregators. According to Auzan (2018), the information society is experiencing a radical reduction in the level of transaction costs and changes in their structure. There is an expansion of exchange options (according to D. North), the appearance of new elements of the spectrum of discrete institutional alternatives and the simplification of collective actions based on cost reduction and the development of sharing (Pazaitis et al., 2017; Oh et al., 2013).

II Sharing Economy as a Special Type of Socio-Economic Relations

Let's imagine the sharing economy as a special type of socio-economic relations that arise on a self-regulating basis between subjects about the collective use of free (excess) resources/goods, as well as implemented using digital technologies and online network services (Table 11.2).

Socio-economic relations of this type arise in industries where there is access to distributed technologies (spare computing cycles, distributed telecommunications, viral communication networks), where other forms of distributed fixed capital are available (for example, in the automobile transport industry), where the design process can be separated from the physical production process, and where financial capital can be distributed (Bauwen, 2005). This type of socio-economic relations is self-regulating and remains outside the system of state administration and regulation.

Table 11.2: Structure and content of socio-economic relations of the sharing economy.

Element of socio-economic relations	Content	Example
An object	Resources and goods (free and excess) for which subjects perform economic actions	Accommodation: Airbnb, HomeAway, OneFineStay, FlipKey. Taxi: Uber, Ola, Grab, Яндекс taxi. Carsharing: Zipcar, Turo, Delimobile. Ridesharing: Blablacar. Things: Avito, Юла, Rentmania, Neighborgoods. Movie: Netflix, Ivi, Amediateka. Knowledge: Skillshare, Coursera (Filimonova et al., 2020).
Subjects (actors)	Main: 1) owners of resources/goods (households, firms) 2) consumers of resources/goods (households, firms) (Botsman and Rogers, 2010) Additional: 3) communities of consumers united by a digital platform where communication and exchange of experience are carried out, they have a direct impact on consumer behavior. 4) the state as a passive subject that provides a solution to the problem of trust	At the initial stage of sharing: The C2C – Consumer-to-Consumer model as an example of implementing the P2P model – peer-to-peer (Bauwen, 2005); At the stage of sharing development: B2C – Business-to-Consumer model – between households and firms. B2B model – Business to Business - between firms (Codagnone et al., 2016; Petropoulos, 2016).
The motives of subjects	Additional opportunities to find temporary work, earn additional income, improve social interaction, and gain access to information resources (Dillahunt and Malone, 2015); the transformation of idle capacity in production resources (Wallsten, 2015).	According to PwC, in 2015, the revenues of European companies belonging to the sharing economy amounted to 4 billion euros, and the total volume of transactions passed through it was 28 billion euros. On average, the platform receives 25% of the cost of the service rendered, and 85% remains with the performer (PwC, 2018).

Table 11.2 (continued)

Element of socio-economic relations	Content	Example
Interaction tool	Digital technologies (digital platforms and online services) allow direct transactions between subjects and provide them with an increase in the value of their own assets (Petropoulos, 2016), they are the most important intermediary in the case of joint economic operations (Codagnon et al., 2016)	Digital platforms perform three roles with different levels of transaction intervention: meeting space, market and matchmaker. Meeting platforms provide communication between users, provide information about each other (profile, reviews, etc.), but are passive (for example, Blablacar). Marketplace platforms facilitate transactions (payment, insurance, etc.) and can become an active party in the event of such disputes (for example, Airbnb or Madpaws). Matchmaker platforms are active intermediaries in deals, match users with similar needs, and interfere in deal negotiations (for example, Uber) (Hafermalz et al., 2016).
Principle of interaction	Shared consumption	A form of practice in which at least two community members engage in direct interaction and use the same units of material goods or services to perform the practice (Huber, 2017).
The mechanism of income distribution	Distribution of payments for the use of a resource/good	Providing consumers with access, but not transferring ownership of the goods; the consumer pays for the temporary experience that the product offers, not for the product itself (Le Vine et al., 2014)
The range of the relationship	Not limited in physical space	In a city, country, or international space

Source: compiled by the authors.

III Economic Benefits and Risks of the Development of Socio-Economic Relations of the Sharing Economy

The absolute benefits of developing socio-economic relations in the sharing economy include: 1) creation of additional added value by subjects (carriers), which leads to an increase in GDP: the size of the sharing economy in the EU-28 in 2016 amounted to 26.5 billion us dollars (about 0.17% of the total GDP of countries) (European Commission, 2018); 2) additional employment of the population (according to McKinsey in 2016, 30% of the working population of the United States had additional part-time jobs, including in connection with sharing) (McKinsey, 2016); 3) reducing the cost of environmental protection due to the reuse of resources and reducing emissions (according to Fremstad, the sharing of housing and things reduces the amount of waste and overall damage to nature (for example, from the construction of new hotels), approximately 700,000 tons of CO_2 emissions were saved during the BlaBlaCar carsharing period (Economist, 2013); 4) the maximum reduction in transaction costs associated with the marketing of free assets due to universal access to the online market, which provides an increase in profitability indicators (for example, in 2018, Airbnb's profit in the United States exceeded the value of the indicator for such hotel chains as Hilton, DoubleTree and Embassy Suites, and actually equaled that of Marriott (Vedomosti, 2019)); 5) digital technologies, as a tool to stimulate demand for goods and services, reduce the value chain (for example, according to the world Bank, 7% of housing rental in 2018 was occupied by peer-to-peer production, and in 2013–2025, the annual growth rate will be 31%) (Worldbank, 2018); 6) reducing the total cost of consumers (due to the lack of direct payment for ownership or related costs, including repairs, maintenance, insurance) and expanding the range of opportunities to implement their experience (Botsman and Rogers, 2010). For example, online sharing platforms increase the cost-effectiveness of vehicle ownership, since personal cars are not used 95% of the time (Sonuparlak, 2011).

The described benefits of the development of this type of socio-economic relations are manifested in the growth of the scale of the sharing economy. So, according to a report by PwC, by 2025, the volume of the sharing economy will approach the mark of 335 billion dollars (PwC, 2018). Based on data collected by Timbro and presented in the form of the Global sharing economy index (monthly traffic data and a full count of active suppliers were collected for 286 services in 213 countries), it can be concluded that the countries with the most developed sharing economy are: Iceland (100), Turks and Caicos Islands (66.9), Malta (58.2), Montenegro (58), New Zealand (52.8), Croatia (52.2), Faroe Islands (49.3), Denmark (45.9), Aruba (43.1), Ireland (41) (Timbro, 2018). At the same time, the World Bank emphasizes that China is the world leader in the sharing economy: in 2018, its volume was estimated at more than 230 billion us dollars, or 1.67% of the country's GDP (WorldBank, 2018). At the same time, the industry-specific share of the economy, according to research by PwC, is that the field of media and entertainment the most covered technology sharing economy by 28%, the rental of housing and real estate 20%, the trade was on a par with the transport services in third place (19%), services (14%), finance (11%), field machinery (10%) (PwC, 2018). However, the risks are probably hidden behind the economic benefits of the sharing economy. We systematize the risks of developing socio-economic relations in the sharing economy, depending on the level of the subject (Table 11.3).

Table 11.3: Risks of development of socio-economic relations of the sharing economy.

Subject	Risk
Consumers of resources/ goods	1) low quality of services due to the fact that the sharing platforms are not actually responsible for the quality of services provided (Cohen, 2015); 2) overestimation of prices for goods and services due to the lack of a systematic approach to pricing (some services set the cost of services centrally (for example, taxi services), while others give users the freedom to negotiate) (Analytical center under the Government of the Russian Federation, 2019); 3) leakage of personal data and the threat of cybercrime due to the openness of information and vulnerability of digital platforms (for example, due to the provision of data during registration in the profile) (Economist, 2013; Karpunina et al., 2019b)
The owners of the resources/benefits	1) economic damage due to physical fraud and theft of resources/ goods (Rbk, 2018); 2) reduced confidence and consumer demand due to poor quality of services (Chica et al., 2017; Hawlitschek et al., 2016); 3) increasing social insecurity of the poor as a result of shifting the tax burden on them (as contractors) (Dostmohammad and Long, 2015); 4) information leaks and the threat of cyber espionage (Karpunina et al., 2019a)
Consumer community	1) reducing the level of trust in sharing platforms; 2) dissatisfaction with the quality of services provided; 3) information boycott for sharing platforms.
State	1) reduced production and slower GDP growth due to increased consumer demand for goods and services in traditional sectors of the economy; 2) violation of the balance in the structure of the economy and the conditions of normal competition (Richardson, 2015); 3) contradictions between the interests of small transactions and the interests of global corporations, the emergence of new forms of inequality and polarization in property relations (Dostmohammad and Long, 2015; Richardson, 2015); 4) shortfall in tax collections due to the lack of institutional bases for taxation of the activities of sharing platforms (Analytical center under the Government of the Russian Federation, 2019).

Source: compiled by the authors.

Conclusion

As a result of the research, it is concluded that the prerequisites for the emergence and development of the sharing economy are changes that were formed in the process of social development and under the influence of digital technologies. They are manifested in the change of concept relations to the property (from the concept of the possession of the concept of using), the change in the model of creation of consumer value (transition from a model of creation of customer value by the firm for the user to model the joint creation of consumer value), in the transformation of the paradigm of commodity relations (concept of commodity exchange was transformed into the concept of a digital exchange), as well as in the transition from the traditional model of economic institutions to retrofit. The authors presented their own vision of the sharing economy as a special type of socio-economic relations that arise on a self-regulating basis between subjects about the collective use of free (excess) resources/goods, implemented using digital technologies and online network services. The authors systematized the economic benefits of developing socio-economic relations of the sharing economy, and also presented a grouping of potential risks for each subject of the sharing economy.

References

Analytical center under the government of the Russian Federation (2019), "The economy of shared consumption as a new economic model", Bulletin on current trends in the world economy, Vol. 47, August 2019.

Auzan, A. (2018), "Cifrovaya ekonomika kak ekonomika sverhnizkih transakcionnyh izderzhek (nabor gipotez)", available at: https://www.econ.msu.ru/sys/raw.php?o=51299&p=attachment (accessed 22 March 2020)

Bauwen, M. (2005),"The Political Economy of Peer Production. 1000 Days of Theory", available at: http://www.ctheory.net/articles.aspx?id=499 (accessed 25 March 2020)

Bell, D. (1986), Social'nye ramki informacionnogo obshchestva, Progress, Moskva.

Bodrunov, S. (2016), "Novoe industrial'noe obshchestvo. Proizvodstvo. Ekonomika. Instituty", Ekonomicheskoe vozrozhdenie Rossii, No. 2 (48), ss. 5–14.

Botsman, R. and Rogers, R. (2010), What's Mine Is Yours: The Rise of Collaborative Consumption, HarperBusiness, NY.

Castels, M. (2000), Informacionnaya epoha: ekonomika, obshchestva i kul'tura, VSHE, Moskva.

Chica, M., Chiong, R. and Adam, M. (2017), "An evolutionary trust game for the sharing economy", in IEEE proceedings Congress on Evolutionary Computation, 2017, pp. 2510–2517.

Codagnone, C., Biagi, F. and Abadie, F. (2016), "The passions and the interests: Unpacking the 'Sharing Economy', available at: http://publications.jrc.ec.europa.eu/repository/ (accessed 12 February 2020)

Cohen, M. and Sundararajan, A. (2015), "Self-regulation and innovation in the peer-to-peer sharing economy", U. Chi. L. Rev. Dialogue, Vol. 82, p. 116.

Dillahunt, T. and Malone, A. (2015), "The promise of the sharing economy among disadvantaged communities", in Human Factors in Computing Systems proceedings of the international conference, April 2015, pp. 2285–2294.

Dostmohammad, S. and Long, J. (2015), Regulating the sharing economy: applying the process for creative destruction, Dalhousie University, Halifax.

Drucker, P. (1969), The Age of Discontinuity; Guidelines to Our Changing Society, Harper and Row, New York.

Economist (2013), "The rise of the sharing economy: On the internet, everything is for hire", available at: https://www.economist.com/news/leaders/21573104-internet-everything-hire-rise-sharing-economy (accessed 17 March 2020)

Ekhart, M. (1991), Duhovnye propovedi i rassuzhdeniya: Reprintnoe vosproizvedenie izdaniya 1912 goda, Politizdat, Moskva.

European Commission (2018), "Study to monitor the economic development of the collaborative economy at sector level in the 28 EU Member States", available at: https://publications.europa.eu/en/publication-detail/-/publication/0cc9aab6-7501-11e8-96334.pdf (accessed 22 March 2020)

Filimonova, N., Kapustina, N., Bezdenezhnykh, V. and Kobiashvili, N. (2020), "Trends in the Sharing Economy: Bibliometric Analysis", in Popkova, E., Sergi, B. (Eds), Digital Economy: Complexity and Variety vs. Rationality.Lecture Notes in Networks and Systems. Vol. 87, Springer Nature, Switzerland.

Fromm, E. (2000), Velichie i ogranichennost' teorii Frejda. Redaktor O. V. Kir'yazev, Izdatel'stvo ACT, Nazran.

Hafermalz, E., Boell, S., Elliot, S., Hovorka, D. and Marjanovic, O. (2016), "Exploring dimensions of sharing economy business models enabled by IS: An Australian study", available at: https://www.semanticscholar.org/paper/Exploring-Dimensions-of-Sharing-Economy-Business-by-Hafermalz-Boell/196c93e394a2915a8e7721dda46d8b395314ede9 (accessed 20 March 2020)

Hawlitschek, F., Teubner, T. and Adam, M. (2016), "Trust in the sharing economy: An experimental framework", in Information Systems proceedings of the International Conference, ICIS in Dublin, Ireland, 2016, pp. 1–14.

Heinrichs, H. (2013), "Sharing economy: A potential new pathway to sustainability", GAIA, Vol. 22 (4), pp. 228–231.

Huber, A. (2017), "Theorising the dynamics of collaborative consumption practices: A comparison of peer-to-peer accommodation and cohousing", Environmental Innovation and Societal Transitions, Vol. 23, pp. 53–69.

Karpunina, E., Gorbunova, O., Moiseev, S. and Cheremisina, T. (2019a), "Resistance is not hopeless . . . regarding the policy of countering information threats of economic security", in 33nd IBIMA proceedings of the International Business Information Management Association Conference in Granada, Spain, 2019, p. 2679–2686.

Karpunina, E., Yurina, E., Kuznetsov, I. and Dubovitski, A. (2019b), "Growth potential and economic security threats in terms of digital economy ecosystem", in 33nd IBIMA proceedings of the International Business Information Management Association Conference in Granada, Spain, 2019, p. 2669–2678.

Kotler, F. (2012), Marketing menedzhment, Piter, Sankt-Petersburg.

Le Vine, S., Zolfaghari A. and Polak, J. (2014), "Carsharing: Evolution, challenges and opportunities (22th ACEA Scientific Advisory Group Report)", available at: https://www.acea.be/uploads/publications/SAG_Report_-_Car_Sharing.pdf (accessed 12 March 2020)

McKinsey (2016), "Independent work: Choice, necessity, and the gig economy: Report", available at: https://www.mckinsey.com/featured-insights/employment-andgrowth/independent-work-choice-necessity-and-thegig-economy (accessed 12 March 2020)

Oh, O., Agrawal, M. and Rao, H. (2013), "Community intelligence and social media services: a rumor theoretic analysis of tweets during social crises", MIS Quarterly, Vol. 37(2), p. 407.

Pazaitis, A., Kostakis, V. and Bauwens, M. (2017), "Digital economy and the rise of open cooperativism: the case of the Enspiral Network Transfer", European Review of Labour and Research, Vol. 23 (2), pp. 177–192.

Petropoulos, G. (2016), "An economic review on the collaborative economy", available at: http://www.europarl.europa.eu/ (accessed 12 February 2020)

Porter, M. (2005), Konkurenciya, Izd. dom Vil'yams, Moskva.

Prahalad, C. and Ramaswamy, V. (2006), The Future of Competition: Co-Creating Unique Value With Customers, Penguin Books, India.

Price Waterhouse Coopers (PwC) (2018), "The Sharing Economy: the new business model", available at: https://www.pwc.de/de/digitale-transformation/share-economy-report-2017.pdf (accessed 28 February 2020)

RBC (2019), "Ekonomika sheringa v 30 cifrah i faktah (Istochniki: Brookings Institute, Global Market Insights, NYSE, NYU, Pew Research, PwC, WEF, RAEK)", available at: https://www.rbc.ru/trends/sharing/5ddbb3279a7947b01be74c19 (accessed 12 March 2020)

RBC (2018), " Doshli do vyruchki: rentabel'nost' shering-ekonomiki – mif ili real'nost'", available at: https://www.rbc.ru/trends/sharing/5de4e5f19a7947f2c0f0d0a7www.rbc.ru/trends/sharing/5de4e5f19a7947f2c0f0d0a7 (accessed 7 March 2020)

Richardson, L. (2015), "Performing the sharing economy", Geoforum, Vol. 67, pp. 121–129.

Roshchina, I. (2005), "Transformaciya ekonomicheskih otnoshenij v usloviyah postindustrial'nogo obshchestva", Vestnik TGPU. Seriya Gumanitarnye nauki (Ekonomika), No. 5 (49), ss. 39–42.

Sonuparlak, I. (2011),"Buzzers" and "Auto-Preneurs" Expand Peer-to-Peer Car-Sharing in France", available at: https://thecityfix.com/blog/buzzers-and-auto-preneurs-expand-peer-to-peer-car-sharing-in-france/ (accessed 5 March 2020)

Tawney, R. (1920), The Acquisitive Society, New York, NY.

Timbro (2018), "Timbro Sharing Economy Index", available at: https://timbro.se/ekonomi/timbro-sharing-economy-index/ (accessed 20 February 2020)

Toffler, E. (1999), Tret'ya volna, ACT, Moskva.

Vedomosti (2019), "Marriott sozdaet biznes po arende zhil'ya dlya konkurencii s Airbnb", available at: https://www.vedomosti.ru/business/news/2019/04/29/800468-marriott-konkurentsii-s-airbnb (accessed 24 March 2020)

Wallsten, S. (2015), "The competitive effects of the sharing economy: How is Uber changing taxis?", available at: https://www.ftc.gov/system/files/documents/public_comments/2015/06/01912-96334.pdf (accessed 15 March 2020)

Worldbank (2018), "Tourism and the Sharing Economy: Policy & Potential of Sustainable Peer-to-Peer Accommodation", available at: http://documents.worldbank.org/curated/en/161471537537641836/Tourism-and-the-Sharing-Economy-Policy-Potential-of-Sustainable-Peer-to-Peer-Accommodation (accessed 14 March 2020)

Aleksei Tebekin, Ekaterina Bogoeva, Andrei Zakharov and Dmitrii Lazarev

12 The Impact of the Fourth Industrial Revolution on the Socio-Economic Development of the World Economy

Introduction

The Fourth Industrial Revolution (Industry 4.0) fueled by avalanche introduction of cyber-physical systems both into the industrial production and services sector certainly has a great impact on various sides of international and national development.

Modern assessments of such an impact are particularly provided in the following works (Shilova, 2018), (Tarasov, 2018), (Isaychenko, 2019), (Dovbiy, 2019), (Shchetinina, 2017), (Burenina, 2018), (Kournitskaya, 2018), (Serebryakova, 2018), (Lisovsky, 2018), (Tebekin, 2018), etc.

Despite various assessments of the impact of Industry 4.0 on the technological prospects for the production development and the expected socio-economic, environmental and political processes in the world, it seems to us that there is a need to make an integrated assessment of the fourth industrial revolution in relation to the world economy and possible changes in the global trends. These circumstances predetermined the relevance of the research.

Methodology

The methodological background of the research is the ideology of the fourth industrial revolution by Klaus Schwab (Schwab, 2018). The technical basis of the research is made up of general scientific methods (including abstraction, analysis, analogy, deduction, idealization, identification, induction, specification, generalization, synthesis, comparison, extrapolation), as well as the SNW analysis.

Aleksei Tebekin, Moscow Institute of International Relations (of the University) of Ministry of Foreign Affairs of Russia, Moscow, Russia
Ekaterina Bogoeva, Russian Customs Academy, Moscow Region, Lyubertsy
Andrei Zakharov, Dmitrii Lazarev, The People's Friendship University of Russia, Moscow, Russia

https://doi.org/10.1515/9783110654486-012

Findings

The impact of the fourth industrial revolution on various sides of world development has been assessed using expert evaluations published in sources covering scientific findings in particular areas.

When interpreting the results of expert evaluations concerning the impact of the fourth industrial revolution on the technological, managerial, economic, social, political, legal and environmental factors of world development, we applied the method of the SNW analysis method spread from the microenvironment to the macroenvironment (Table 12.1).

Table 12.1: Interpretation of expert evaluations concerning the impact of the fourth industrial revolution on the technological, managerial, economic, social, political, legal and environmental factors of world development using the SNW analysis method.

No.	Factors of world development	The nature of the impact of the fourth industrial revolution		
		Positive (S)	Neutral (N)	Negative (W)
1	Technological factors	S, S		
2	Managerial factors	S, S		
3	Economic factors	S	N	
4	Social factors	S		W
5	Political Factors		N	W
6	Legal factors	S		W
7	Environmental factors	S	N	
8	Integrated assessment	S = 8	N = 3	W = 3

It should be noted that we have taken into account various opinions and expert justifications in making an integrated assessment. The differences in expert opinions are presented in Table 12.1.

As follows from the evaluations presented in Table 12.1, the unanimity of experts is associated with the solely positive influence of the fourth industrial revolution on the technological factors of development.

All the experts are sure that the twelve basic technologies of Industry 4.0, demonstrating broader opportunities of digital technologies (including blockchain and distributed ledger technologies, the Internet of things and new computing technologies, for example, quantum computing technologies), possibilities of the physical world transformations (including technologies of additive manufacturing and multidimensional printing, artificial intelligence and robotics, as well as technologies for the generation of new materials), changing human capabilities (including biotechnology, virtual and augmented reality and neural network technologies), environmental integration capabilities (including geoengineering technologies, space technologies, and technologies for the development of eco-friendly renewable energy sources) will act as a driver to ensure the accelerated development of a whole range of other post-industrial technologies in the 21st century.

The experts are unanimous also in the positive impact of Industry 4.0 on the managerial factors of world development (Tebekin, 2018).

As for the impact on the economic factors, then the experts have different opinions.

Some of them argue that the fourth industrial revolution will definitely ensure the growth of the global economy. Others believe that despite exponential growth and investments in new technologies and the technological progress itself accompanying the fourth industrial revolution global economic growth slows down within a saturated market and may even stop.

Along with that, it should be marked that the second group of experts mainly take into account the indicators of economic development in the G7 countries (Great Britain, Germany, Italy, Canada, the United States of America, France, Japan). However, the first group of experts considers the high rates of economic development in some countries of the E7 (Brazil, India, Indonesia, China, Mexico, Russia, Turkey).

The standpoint of experts stating the positive impact of Industry 4.0 on the growth of the world economy is supported by the following facts.

Firstly, according to data of the International Monetary Fund on total GDP calculated using purchasing power parity, E7 countries surpassed G7 countries as far back as 2013 ($ 37.8 trillion versus $ 34.5 trillion) (Samofalova O., 2014).

Secondly, according to PricewaterhouseCoopers forecasts, in 2050 E7 countries will outrun G7 countries in terms of bank assets (PwC, 2015).

At the same time, the slowdown of growth rates can be objectively explained by higher competition between the "old" and "new" economic leaders, which resulted in trade wars (DiChristopher, 2018).

Besides, we can't but take into consideration the patterns of economic activity within technological waves, which under N. Kondrat'ev large cycles of economic activity, suppose the outbreak of a global economic crisis during a shift to the sixth technological paradigm in 2020 (Tebekin, 2018). These patterns refute the statement of Klaus Schwab premised on the development indicators of the G7 countries that digital technologies have already made a pioneering contribution to the world economy. Nowadays, they don't give a boost to productivity (Schwab, 2018).

Also, we cannot ignore a widening gap between the leading countries of the world with an advanced production-technology base developing thanks to new digital technologies (as a substructure) and underdeveloped countries where digital technologies cannot be applied in many economic sectors due to insufficient competitive production-technology base in terms of Industry 4.0.

Investigating the social factors of the world economy in relation to Industry 4.0, some experts are on opposite ends. One group of experts reasonably believes that digital technologies will have a favorable impact on the social development, also using the concepts of "smart home", "smart city" (Tebekin, 2019), etc. Another group of experts not without good reason believes that modern technological development causing an economic gap between countries and different social classes will surely enhance social stratification (Galkin, 2016).

As for the political factors of world development, some experts suppose that the fourth industrial revolution will not have a great impact on world political processes (Tebekin, 2018), but others are convinced that the erosion of the middle class due to social stratification will very likely upset the balance between political systems (Galkin, 2016).

Experts are concerned about the impact of Industry 4.0 on the legal factors of world development. If the technologies can speed up technical legal procedures, then due to the acceleration of scientific and technological progress, the gaps between the socio-economic and political processes taking place "de facto" and fixed "de jure" will deepen and even lead to irreversible effects.

Exploring environmental factors in the wake of the fourth industrial revolution, we should stress that one group of experts believes that the application of energy- and resource-saving technologies will have a positive impact on the environmental situation on the planet. Other experts reckon that dominating transnational corporations moving hazardous productions to less developed countries will not care about the high costs for ensuring environmentally friendly production (Tebekin, 2014).

Conclusion

In general, an integrated assessment of the impact of the fourth industrial revolution on various factors of world development showed that Industry 4.0 will have a positive impact on the technological and managerial factors. Some positive impacts will cover economic, social, legal and environmental factors of world development. At the same time, experts emphasize that the social, political and legal factors of the world can be negatively affected. In this situation, the experts are highly concerned about the negative impact of the fourth industrial revolution on the world political factors.

References

Burenina I.V., Gaifullina M.M., Saifullina S.F. Socio-economic transformations associated with the implementation of projects on design and introduction of Industry 4.0 technologies // *Bulletin of the Eurasian Science*, 5, 2018. PP. 1–15.

Dovbiy I.P., Ionova N.V., Dovbiy N.S. Fourth industrial revolution (investment-financial and staffing issues). *Bulletin of the South Ural State University, Ser. Economics and Management, 13 (1)*, 2019. PP. 120–131.

Galkin S. Industrial Revolution 4.0. Reed Media. URL:https://web.archive.org.

Isaichenkova V.V. Ensuring the competitiveness of an industrial enterprise through comprehensive performance analysis. *Leadership and Management, 6 (3)*, 2019. PP. 177–188.

Kurnitskaya K.Yu. The fourth industrial revolution and innovative trends in the global and national economies. Economic Science and Practice: Proceedings of the 6th Intern. scientific conf. Chita, 2018. PP. 16–19.

Lisovsky A.L. Optimization of business processes for the transition to sustainable development in the conditions of the fourth industrial revolution. Strategic Decisions and Risk Management, 4, 2018. PP. 10–19.

Samofalova O. Seven new. The group of developing countries, including Russia, is ahead of G7 by several indicators. Available at: https://vz.ru/economy/2014/10/9/709649.html

Schwab K. Technologies of the fourth industrial revolution. Eksmo, 2018. 320 p.

Serebryakova N.S., Petrikov A.V. The design and organization principles of innovative infrastructure performance in the context of the Industry 4.0. Proceedings of the Voronezh State University of Engineering Technologies, 80 (4), 2018. PP. 384–387.

Shchetinina N.Yu. Industry 4.0: practical issues in Russian conditions. *Models, Systems, Networks in Economics*, Technology, *Nature and Society, 1 (21)*, 2017. PP. 75–84.

Shilova E.V., D'yakov A.R. On the phenomenon of the fourth industrial revolution and its impact on the economy and management. Bulletin of the Prikamsk *Social Institute, 3 (81)*, 2018. PP. 86–95.

Tarasov I.V. Industry 4.0: Technologies and their impact on the productivity of industrial companies. *Strategic Decisions and Risk Management, 2*, 2018. PP. 62–69.

Tebekin A.V. Analysis of crises from the standpoint of economic theory. // *Journal of Economic Research, 4 (12)*, 2018. PP. 3–9.

Tebekin A.V., Bozrov A.M. International investment activities of transnational corporations with regard to globalization trends and regional features. Transport Industry of Russia, 4, 2014. PP. 35–37.

Tebekin A.V., Egorova A.A. Solving social urban problems with the help of Smart City technologies: challenges and prospects // *Journal of Sociological Research, 4*, 2019. PP. 32–46.

Tebekin A.V. Development trends and prospects of political analysis techniques for international relations and global and regional systems. *Journal of Political Studies, 2 (3)*, 2018. PP. 125–134.

Tebekin A.V., Petrov V.S. Analyzing the transformation of the main provisions on effective management in shifting of society from industrial towards the post-industrial economy. *Business Strategies, 12*, 2018. PP.3–12.

Tom DiChristopher. (2018) / Trump's solar tariffs could put the brakes on rapid job growth in renewable energy. Available at: https://www.cnbc.com/2018/01/23/trumps-solar-tariffs-could-slow-down-rapid-renewable-job-growth.html

PricewaterhouseCoopers: Forecast of the global economy from 2015 to 2050. (2015). Available at: https://gtmarket.ru/news/2015/02/11/7089

Part IV: **High Technologies and State Management in Industry 4.0**

Kamila V. Kudryavtseva, Moisey A. Skliar, Lidiya R. Vakhitova and Natalya A. Shapiro

13 Economics of Industry 4.0 in the Political Economy Paradigm

Introduction

Based on the fact that the digitalization of the economy, as noted by Negroponte (Negroponte, 1995), is a factor in the global transformation of the modern world, then its influence should be manifested not only in changing the economy, but also in the system of social relations of the national economy (Atrokhova and Alpidovskaya, 2019). The work of C. Schwab "The Fourth Industrial Revolution" (Schwab, 2018) introduced the term "industry 4.0" into the scientific discourse and presented a comprehensive description of digital technologies. The economy, according to industry 4.0, is changing based on a new type of industrial production (Ilin et al. 2019), using big data, automation, blockchain (Plotnikov and Kuznetsova, 2018), the Internet of things, artificial intelligence, smart platform networks (Kudriavtseva, 2019); (Skliar and Kudriavtseva, 2019) and more. But, although there is a unanimous opinion on the topic of technological and economic changes in society, there is not only agreement on social relations themselves, but also unambiguous ideas about these changes, despite the fact that the topic of the digital economy has been actively discussed in Russia for more than 10 years (Vakhitova, 2007), (Dyatlov ed., 2007), (Ayrapetova et al., 2010). Among domestic political economists, discussions of alarmist warnings (Kulkov, 2019), (Nedzvetskaya, 2019), which can accompany the transformation of relations in industry 4.0, are especially notable. The reason for this may be a list of the 21st technological change arising from the technologies of industry 4.0, the approximate date of their appearance on the market, as well as turning points that entail both positive and negative consequences for society, contained in the final part the work of K. Schwab. Such accents are explained by the complexity of the transformations, some cautiously perceive the new, others hardly part with the old: "accept and deal" or "refuse and lose". But the alter-native, as C. Schwab writes, is not the point (Schwab, 2018), it consists in how to use the enormous advantages of the technological revolution to make life better.

Kamila V. Kudryavtseva, Moisey A. Skliar, Lidiya R. Vakhitova, Natalya A. Shapiro, Herzen State Pedagogical University of Russia, St. Petersburg, Russia

https://doi.org/10.1515/9783110654486-013

Methodology

The methodological tool of this study is the integration of conceptual capabilities (Malyshev et al., 2006) of two fundamental general scientific concepts: "paradigm" and "disruption". The "Paradigm" allows us to understand the differences between political economy and economic theory, as well as the internal features of the concepts of political economy, reflecting the transformation of society in the conditions of scientific and technological progress, among themselves.

Economic science, which is conceptually immersed in the study of social aspects, is political economy. Its task is to show the main institutional actors, explain their behavior and functions in the processes of production, distribution, exchange and consumption of "wealth of peoples" in the specific historical circumstances of economic interaction (Shapiro, 2013). When an economy is considered in isolation from issues of social interaction, it becomes a science denoted by the term "economics" or economic theory. Economic theory includes a certain core of "pure" market theory (the interaction of demand, supply and equilibrium), and explores a wide field of applied issues of the behavior of individualized market entities, both on the demand and supply sides. "Economics" (or economic theory) is a wider and more abstract science than is the case within the narrower term "Political Economy" (political economy). This is how the distinction between political economy and economics was interpreted by A. Marshall (1993), the author of the term economics and one of the founders of microeconomic theory.

The "disruption" or hypothesis of disruptive innovation lies at the heart of the metaphysical picture of impending reality that industry 4.0 forms. The disruptive innovation hypothesis was put forward by Joseph L. Bauer and Clayton M. Christensen at the end of the 20th century (Bower and Christensen, 1995). After 20 years, K. Christensen again turned to this idea due to the fact that the term "disruption" has become too often used in the free sense to describe any situation in which the industry is shaking and previously successful employees cease to be so. But such widespread use distorts the original meaning (Christensen et al., 2015).

In this study, "disruption" is seen as unpredictable change.

Results

Changes in the technological parameters of production or industries are reflected in different ways in economic theory and political economy. If economic theory fixes changes in the behavior of economic entities (sellers and buyers or firms and households) through various kinds of effects, traps, expectations, etc., then in political economy the institutional status of the economically significant social entities themselves and the distribution of responsibility between them for the results are important economic development (selected in responsibility for progress), as well as the form of manifestation of the result.

So, the three preceding industries, successively replacing each other throughout the twentieth century: the industry of large machine production or the "chimney industry", then the mass consumption industry and the information society industry – were reflected in the corresponding models of social interaction and canonized in educational publications.

Large-scale machine production at the beginning of the twentieth century was considered in the framework of the social, conflict interaction of wage workers and capitalists. Capital and capitalists are responsible for the progress of society. The wealth of nations was represented by a mass of goods produced on the basis of a large machine industry.

Mass consumption, formed by the middle of the twentieth century, was based on mass production lines of mass production of quality goods and services, including new goods and services previously unknown to the mass consumer. The main social and consumer link in social development began to consider nuclear families as the basis of the middle class. The wealth of the nation is the flow of sustainable consumer demand, providing dynamic markets, and, consequently, economic growth. The scale of mass production has allowed globalization of consumption on a supranational scale. The economy or GDP began to grow at the expense of the services sector (post-industrial society).

The information society of the 1980–1990s appeared as a result of the accumulation and development of scientific knowledge that created intelligent technologies. A society of mass individuality was created. The main form of nation's wealth is becoming the intellect – the intellectual potential of people (human capital), territories, intellectual property, aimed at ensuring the development of innovations, the effective use of the capabilities of information production, the development of means of adaptation to changes. The main character in society is an entrepreneur who is able to generate and introduce innovations. The economy is developing at the expense of non-profit public enterprises (social entrepreneurship).

The models of reality in the three indicated industries qualitatively differed in the institutional structure of the economy, social differentiation, and responsibility for progress. With a certain degree of conventionality, we can say: about the class of capitalists (Marxism and Soviet experience sought to show that the progressive development of society is achieved, first of all, thanks to the working class and without the class of owners of capital), about the state, which provided general prosperity, and about elites aimed to create institutions that foster innovation and entrepreneurship. Theories reflecting social transformations based on scientific and technological progress have themselves transformed very dynamically. This is evidenced by the history of theories of transformations presented in the works of such authors as J.K. Galbraith, W. Rostow, P. Drucker, D. Bell, A. Toffler (Khudokormov, ed., 1998). Each of them regularly corrected the conceptual vision of how progress is being realized, because of what and why social conflicts arise, why the intended goals are not achieved and so on. But, the fact of the changes being introduced does not at all indicate subjective or cognitive difficulties, it only shows the dynamism of the changes and the difficulty of predicting the consequences that the future carries in itself, determined by the achievements of scientific and technological progress.

The advantage of the concept of industry 4.0 is that the fact of the complex predictability of changes is laid at its foundation. The metaphysical picture of impending reality is based on the development of the hypothesis of disruptive innovations (disruptive innovations are in Russian vocabulary), K. Schwab believes: "The scale and scope of the changes explain the acuteness of perception of disruptive innovations at the moment" (Schwab, 2018). In this case, disruption is understood as breaking the usual notions, which are manifested in unpredictable effects, breakthroughs, organizations, relationships, and so on.

Digitalization technology is the material premise of this hypothesis. Digitalization is a qualitatively different technological standard with respect to all previous technologies called analogue (Tapscott, 1995), and generates a multifaceted and deep interdependence of the world. It forms the hyperconnection of the future world and collaborative innovation. Hyperconnection makes it difficult to separate the effect of one particular step or action from another, affects the relevance of traditional economic assessments of the results of social development in macroeconomics: GDP, investment, savings, consumption, employment, inflation, etc. Digitalization itself is capable of synthesizing all new technologies adapted to the conditions, goals and objectives of reality, which leads to unprecedented changes in paradigms representing the economy, business, society for each individual person, and society as a whole.

Any innovations, including collaborative ones, reinforce the disruptiveness of the consequences, because this is something new, unknown and kept secret from competitors to enter the market, not fully understood. With hyperconnection of innovation, there is a factor enhancing unpredictability or disruption.

The disruption hypothesis differs from the traditional uncertainty and risk approach. The difference is that the hypothesis of uncertainty implies the possibility of its assessment, and the correction of risks in the actions taken. The risk can be reduced, but can be avoided (as an extreme case, do not take the intended action). The hypothesis of disruption excludes the regularity as such, and the manifestation of disruption is possible even in case of refusal of any active actions, i.e. nothing can be done, and changes can be, as a result of hyperconnection.

In addition to the systemic factors of digitalization: the introduction of platforms, the economy on demand, the sharing economy, the reduction of the marginal cost level and the reproduction of "non-competing" goods and services or the use of digital platforms, there are structural factors: demography, finance, politics, culture (Bowers et al., 2017), which also work towards increasing the unpredictability of change (Anderson and Wladawsky-Berger, 2016). Actions at the same time they reduce the unambiguity of the development trends of the labor market and the training system for professions.

We cite well-known economic facts as an illustration of the disruptive manifestations of both positive and negative. So, Instagram or WhatsApp did not actually need financing to start, which changed the traditional idea of the role of capital in scaling a business. And the international project on the internationalization of education was originally supposed to be based on the values of cooperation, partnership, exchange, mutual benefit and mobilization of forces, and, according to its initiators, will turn into a process associated mainly with concepts such as strong competition, commercialization, prudence and status (Knight and Wit, 2018).

K. Schwab, like many other experts, believes that the main factor of development will be not material, but human capital, embodied in human resources. A shortage of competent personnel, and not a lack of capital, will be a constraining constraint on innovation, competitiveness and growth. However, we note that the practices of all previous industries, as well as their concepts, could not overcome the factor of the influence of the size of money capital on the level of income and welfare of citizens, affect the trend of increasing income differentiation and concentration of wealth. This indicates that the scientific and technological revolution does not automatically lead to a change in relations in society, it creates only the prerequisites for choosing a variety of options.

Conclusion

A positive assessment of the hypothesis of the disruption of industry 4.0 within the framework of the political economic paradigm is that it is moving away from constructing society as a kind of complete model or a point of a happy state but indicates the real need for control on the part of society over the changes taking place. Relative constraints on the development of such a society at any given moment in time are its energy potential (for the development of ICT) and the creative potential of the population (for the development of innovation).

Artificial intelligence and big data will help save time spent on making decisions and choosing options, suggest the consequences of possible actions taken by each individual or any institutional entity, including the state. We can say that the humanism of human behavior acquires a technological basis in calculating the consequences, which is, of course, a new field for the development of measures, norms and rules in society. Virtual experiments will help to reject or accept development options considering moral criteria, but the choice of development criteria and models will be a product of human intellect and morality, reflect the results of the interaction of social forces, i.e. everything is as it was before.

Thus, the political and economic paradigm of industry 4.0 is an open system, the filling of which depends on the specific goals and objectives of society, the validity and calculation of social choice.

References

Alpidovskaya, Marina L., Sokolov Dmitry P. (2019) The Comprehensive Plot of Socioeconomic Development of Russia in the Conditions of the Technological Revolution of the Age of Globalization. Tech, Smart Cities, and Regional Development in Contemporary Russia. UK, Emerald.

Anderson, L., Wladawsky-Berger, I. (2016). «The 4 Things It Takes to Succeed in the Digital Economy», 2016. URL:https://hbr.org/2016/03/the-4-things-it-takes-to-succeed-in-the-digital-economy(accessed 24 March 2018)

Atrokhova, A.N., Alpidovskaya, M.L. (2019). «The effect of Digitalization of the Economy on the Nationa», Economic Processes, Self-management, Vol. 2, No 3 (116), pp. 28–31.

Ayrapetova, A.G. (et al.) (2010). «The state and the market. Mechanisms and regulatory methods in the context of the transition to innovative development», Collective monograph: St. Petersburg, Vol.1, 394p.

Bower, J.L, Christensen, Cl.M. (1995). «Disruptive Technologies: Catching the Wave», Harvard Business Review, January–February. https://hbr.org/1995/01/disruptive-technologies-catching-the-wave?_ga=2.239673027.1137346222.1581149828-578726034.1581149828 (accessed 08 February 2020).

Bowers, M.R., Hall, J.R., Srinivasan, M.M. (2017). «Organizational culture and leadership style: The missing combination for selecting the right leader for effective crisis management», Business Horizons, Vol. 60. Issue 4, July–August, pp.551–563. URL:http://www.sciencedirect.com/journal/business-horizons/vol/60/issue/4 (accessed 24 March 2018).

Christensen, Cl.M., Raynor, M.E., McDonald, R. (2015). «What Is Disruptive Innovation», Harvard Business Review, December, pp. 44–53.

Dyatlov S.A. (ed.) (2007). «The state and the market: a new quality of interaction in the information and network economy», Collective monograph, In 2 Vol. St. Petersburg: Asteron, 396p.

Ilin, I.V. (et al.) (2019). «List of requirements for the architecture of the digital space of Russian business to the technologies providing its realization List of requirements for the architecture of the digital space of Russian business to the technologies providing its realization». Scientific journal NRU ITMO Series "Economics and Environmental Management", No 4 (39), c.72–79, DOI 10.17586/2310-1172-2019-12-4-72-79

Khudokormov, A.G. (ed) (1998). «History of economic studies: (current stage)», M.,: INFA-M, pp. 480–515.

Knight, J., Wit, H. (2018). «Internationalization of Higher Education: Past and Future», International Higher Education, No 95, p. 7.

Kudriavtseva, K.V. (2019). «Digital transformation of retail trade», Problems of Modern Economics, No 2 (70), pp. 182–184.

Kulkov, V.M. (2019). «Digital fetishism», Philosophy of economy, No 4 (124), pp. 109–116.

Malyshev, A.A. (et al.) (2006). «The components of the integration management of economic education», Bulletin of the International Academy of Refrigeration, No 1, pp. 6–9.

Marshall, A. (1993). «Principles of economic», Moscow, Publishing Group "Progress" "Univers", Part I, 414 p.

Nedzvetskaya, N.P. (2019). « The Social Age of the Digital Age», Philosophy of Economy, No. 4 (124), pp. 219–225.

Negroponte, N. (1996). «Being Digital», Vintage Books, ISBN 0-679-76290-6, 255p.

Plotnikov, V., Kuznetsova, V. (2018). «The prospects for the use of digital technology "blockchain" in the pharmaceutical market», In the Collection: MATEC Web of Conferences, p. 02029.

Shapiro, N.A. (2013). «Intentions for re-actualization of political economy: a critical look», Problems of Modern Economics, No 2 (46), pp. 41–43.

Skliar, M.A., Kudriavtseva, K.V. (2019) «Digitalization: Main Directions, Advantages and Risks», Economic revival of Russia, No 3 (61), pp. 103–114.

Schwab, Kl. (2018). «The Fourth Industrial Revolution», Moscow: Publishing House "E", 208 p.

Tapscott, D. (1995). «The Digital Economy: Promise and Peril in the Age of Networked Intelligence», NewYork: McGrawHill, 400p.

Vakhitova, L.R. (2007). «The Impact of Information Technology on Economic Growth and Productivity», Bulletin of the Russian State Pedagogical University A.I. Herzen, Vol. 11, No 32, pp.54–58.

Marianna S. Santalova, Igor L. Surat, Dariko K. Balakhanova,
Irina V. Soklakova and Vladimir I. Surat

14 Features of the Company's Organizational Culture

Introduction

The relevance of the topic chosen for research is primarily due to the increased requirements for the level of the individual's psychological state and how it will manifest itself in the workplace. Improving the social and psychological climate of the team is the way to effectively implement its organizational culture.

The most significant elements of culture are recognized: values, mission, company goals, codes and standards of conduct, traditions and rituals.

The significance of the study of the socio-psychological (organizational) climate is determined by the fact that it can act as a factor in the effectiveness of the organizational culture of the workforce.

Methodology

A great contribution to the study of organizational culture was made by such scientists as E. Shane, G, Hofsted, S Hundy, W. Ouchi, C. Barnard, Santalova M.S. (Santalova - M.S. et al., 2006, 2014, 2017, 2019), Bodrova M.I. (Bodrova M.I., 2019), (Guskova N.D., 2014) and others. Theories of organizational culture were developed by K. Cameron, R. Quinn, Kayl I.I. (Kayl I.I et al., 2017) and others. The problem of the socio-psychological climate as a component of organizational culture is quite well studied in management. This phenomenon was considered in the works of Volkova I.P., Kuzmina E.S., Parygina B.D., Platonova K.K., Shepeli V.M., Sventsitsky Yu. A.L., Spivak V.A., Kachan P.A. (Kachan P.A.,2010), Rodionov E.V. (Rodionov E.V. ey al.,2012) and others. Methods for measuring the socio-psycholoigical climate in the team were used by J. p. Moreno, A. S. Mykhailiuk, L. Y. Sharito, Yukhneva E.A. and others.

Note: The research results were discussed in the process of holding round tables of the Management department of the Plekhanov Russian University of Economics and Moscow Economic Institute, Russia.

Marianna S. Santalova, Igor L. Surat, Irina V. Soklakova, Vladimir I. Surat, Moscow Economic Institute, Moscow, Russia
Dariko K. Balakhanova, Plekhanov Russian University of Economics, Moscow, Russia

https://doi.org/10.1515/9783110654486-014

Results

The organizational culture of the company affects the labor behavior of workers as a set of actions and actions that reflect the internal attitude of people to the conditions, content and results of activities, but lend themselves to objective measurement, assessment and external impact.

For an integral characteristic, various similar definitions in terms of content are used in the literature "socio-psychological climate," "moral climate", "psychological climate", "organizational climate".

The socio-psychological climate is always a reflected, subjective entity, which, although closely related to the objective conditions of life of a group of people, is not identical to them. So, for example, the nature of the relationship in the group acts as a factor affecting the climate, but is not an element of the climate – the climate will be the perception of these relationships by the members of the group. The essential characteristic of the climate is the overall emotional and psychological mood. Climate is understood as the mood of a group of people that determines their labor behavior.

In his work "Inductive Measurement of the Psychological Climate," Coase (Table 14.1) analyzed a significant number of theoretical works and identified more than 80 different dimensions of the organizational climate proposed by various researchers. Having categorized the proposed measurements, he identified 6 dimensions by which the organizational climate is different.

Table 14.1: Measurement of the organizational climate in the team (Coase).

Parameters of the organizational climate	Parameter summary
Autonomy	Perception of one's independence in relation to work procedures, goals and priorities of work.
Cohesion	Perception of closeness in business units, willingness to help, including material resources.
The trust	Perception of freedom of communication with more status colleagues, including on significant or personal issues, confidence that information will not be used to the detriment.
Pressure	Perception of the degree of exactingness in relation to the quality and timing of tasks.
Support and recognition by the leader	Perceptions of managers' tolerance of employee behavior, including a willingness to allow employees to learn from mistakes without fear of reprisal. The perception that employee contributions will be appreciated.
Honesty	The perception that all organizational practices are fair, nonrandom and constant.

Exval E. proposed to evaluate the organizational climate by the following ten parameters: motivation, liberty, support for ideas, openness, dynamism, humor, diversity of opinion, conflict, risk taking, time spent on ideas.

As you can see, many parameters are present in all proposed classifications, which confirms the idea of climate as an integral socio-psychological phenomenon and a component of the organizational culture of the company. Despite the difference in terminology, most researchers agree that a favorable socio-psychological climate is a condition for increasing labor productivity and job satisfaction.

Consider in Table 14.2 the parameters, necessary for measuring the socio-psychological climate in the labor collective, proposed by Zammat and Krackover.

Table 14.2: Measurement of the socio-psychological (organizational) climate in the labor collective (Zammato and Krakovera).

Corporate (corporate spirit)	In organizations with a high level of corporate spirit, employees are self-confident and enthusiastic about what is happening in the company. Important is the "honor of the uniform." In organizations with a low level of corporate spirit, employees are sometimes apathetic, do not feel that they have a goal, are not sure about the future
Conflict	In organizations with a high level of conflict, there is a constant confrontation of forces, goals and beliefs, friction and contradiction. In organizations with a low level of conflict, harmony reigns in goals and beliefs, which inspires employees on work and cooperation.
The trust	In organizations with a high level of trust, employees are open and truthful, they are confident in the safety of information shared with colleagues. In organizations with a low level of trust, employees are closed, careful, and insincere. An atmosphere of anxiety and danger reigns
Remuneration	The organization has a high level of remuneration if the remuneration is perceived as fair and reasonable, without warning or favoritism. The organization is unfair in its remuneration, if the basis for the remuneration is "favoritism", personal preferences, criteria that are not related to work.
Resistance to change	The organization is characterized by high resistance to changes, when employees are inertial and strive to "do tomorrow as they did today." In companies with low resistance to change, employees accept changes as normal and enjoy the fact that tomorrow may be different.
Belief in a Leader	Belief in a leader is high. When employees believe in the leadership of a leader, there is a sense of respect, inspiration, decision-making and action. Belief in a leader is low when employees do not respect and accept the legitimacy of authority.
Shifting responsibility	With a high level of responsibility shifting, employees believe that responsibility for actions rests with others – top management, colleagues, or someone else. At a low level – employees believe that they themselves are responsible for the failure of their actions.

As you can see, these parameters allow you to evaluate the organizational culture sellers, warehouse workers, managers (based on the parameters of measuring the socio-psychological climate of Zammato, Coase and Exval). The assessment was made on a five-point scale. The values determined are as follows: 5 points – a high level of the parameter; 4 points – a fairly high level of the parameter; 3 points – the average level of the parameter; 2 points – low level of the parameter; 1 point – a very low level parameter. As a result of the study, the following data were obtained, which present in Tables 14.3–14.5.

After evaluating the answers received, we can draw the following conclusions: – not a single parameter was evaluated at 5 points (this was predictable, since it is difficult for the company to achieve such a socio-psychological climate). Most of the parameters received 4 and 3 points. In general, these are good indicators, but it is worth paying attention to certain points: – a good indicator is that few respondents rated the parameters at 2 points. Only a few people rated the parameters at 1 point. We drew attention to the parameters that received a rather low rating from many respondents – 3 and 2 points, and it is to improve these parameters that we offer the following.

These parameters include:

1. Autonomy – 66 people rated this parameter by 3 points (this is the average level of the parameter) and 8 by 2 points (low). Here it is recommended to give employees more independence in their work, as this factor can increase the level of perception of their own autonomy, but also the level of responsibility and independence of employees.
2. Cohesion – 33 people rated this parameter at 4 points (high level) and 40 people at 3. To increase the sense of unity (the most important parameter of the socio-psychological climate), we recommend holding various corporate events: congratulations on holidays; field trips; competition between collectives of different stores. In our opinion, a fair attitude of management towards employees also contributes to a sense of unity. Of great importance is the development of a sense of belonging to the organization – employees must be aware of the goals of the enterprise, have information and be involved as active participants in achieving its success. The management should encourage mutual support, mutual assistance in the team.
3. Confidence – 21 people rated this parameter 4 points, 52 people 3 points.
The following measures can be taken to increase confidence in the work team of a company: fair treatment of all employees; providing management with effective feedback (sufficient mutual awareness of significant issues); good communicative qualities of the leader contribute to the development of a sense of trust (attentive attitude to subordinates, politeness, friendly attitude, tact); constantly increasing the level of competence of the leader.
4. Support and recognition by the leader – 4 points for this parameter were put by 19 people, 3 points – 31, and 21 people put 2 points. This is an alarming result, which indicates a rather low level of expectation of support and recognition by the leader.

Table 14.3: Measurement of the parameters of the organizational climate of the company (Kouz).

Parameters of organizational climate	Brief description of the parameter of the organizational climate	The number of people who rated the parameters by 5 points	The number of people who rated the parameters by 4 points	The number of people who rated the parameters by 3 points	The number of people who rated the parameters by 2 points	The number of people who rated the parameters by 1 point
Autonomy	Perception of one's own independence in relation to work procedures, goals and activity priorities.	0	5	66	8	1
Cohesion	The perception of intimacy in opganizatsionnyh installations. willingness to help, including material resources	0	33	40	7	0
The trust	Perception of freedom of communication with more status colleagues, including on significant or personal issues, confidence that information will not be used to the detriment.	0	21	52	7	0
Pressure	Perception of the degree of exactingness in relation to the quality and timing of tasks.	0	23	34	3	0
Support and recognition by the leader	Perceptions of managers' tolerance of employee behavior, including a willingness to allow employees to learn from mistakes without fear of reprisal. Perception that employee contributions will be appreciated	0	19	31	20	10
Honesty	The perception that all organizational practices are fair, nonrandom and constant.	0	38	26	12	4

Table 14.4: Measurement of the parameters of the organizational climate in the work team (Exval).

№	Parameters of the organizationalclimate	Number of people evaluating parameters 5 points	Number of people evaluating parameters 4 points	Number of people evaluating parameters 3 points	Number of people evaluating parameters 2 points	Number of people evaluating parameters 1 point
1	Motivation	0	8	58	15	1
2	Ideas support	0	23	40	7	0
3	Openness	0	13	36	30	1
4	Dynamism	0	28	50	2	0
5	Humor	0	43	37	0	0
6	Diversity of opinions	0	34	30	12	2
7	Conflicts	0	48	30	2	0
8	Risk taking	0	35	37	6	2
9	Time spent on an idea	0	29	42	9	0

Table 14.5: Measurement of the socio-psychological climate in the labor collective according to Zammato and Krackover.

The parameter of the organizational climate	Brief description of the parameter	Number of people, rated 5 points	Number of people, rated 4 points	Number of people, rated 3 points	Number of people, rated 2 points	Number of people, rated 1 point
Corporate spirit	In organizations with a high level of corporate spirit, employees are confident and enthusiastic about what is happening in the company. Important is the "honor of the uniform."	4	38	38	0	0
Conflict	In organizations with a high level of conflict, constant confrontation between forces, goals and beliefs, friction and contradictions between employees.	0	43	37	0	0

Table 14.5 (continued)

The parameter of the organizational climate	Brief description of the parameter	Number of people, rated 5 points	Number of people, rated 4 points	Number of people, rated 3 points	Number of people, rated 2 points	Number of people, rated 1 point
The trust	In organizations with a high level of trust, employees are open and truthful, employees are confident in the safety of information shared with other employees.	0	24	363	20	0
Remuneration	The organization has a high level of remuneration, if the remuneration is perceived as fair and reasonable, without warning or favoritism.	0	16	42	22	0
Resistance to changes	The organization is characterized by high resistance to changes when employees are inertial and strive to "do tomorrow, just as they do today".	0	47	30	3	0
Belief in a Leader	Belief in a leader is high. When employees believe in his leadership, there is a sense of respect, inspiration, decision-making and action.	0	49	31	0	0
Shifting responsibility	With a high level of responsibility shifting, employees believe that responsibility for actions rests with others – top management, colleagues, or someone else.	0	64	16	0	0

To increase the perception of this parameter, we recommend the following: fair criticism of the employee: criticism should be carried out in private with his subordinate, only in essence, the employee should know exactly what they criticize him for, must be in the form of criticism polite and correct; the leader must be aware of the success of employees (feedback); the manager should not skimp on the employee's praise for the results achieved, the most effective praise is publish the attitude to the staff is friendly, fair; leadership style is democratic.

Let us evaluate the data obtained: The parameter motivation – 14 people rated 4 points, 42 respondents rated 3 points and 21 respondents 2 points. Humor – 4 points were given by 37 respondents and 43 respondents 3 points. This is a good indicator of satisfactory compatibility of team members. Openness. According to this parameter, 13 respondents gave a score of 4 points, 36 – 3 points and 30 – 2 points. This parameter is similar to the Coase parameter – trust. Conflicts. According to this parameter, a good result was obtained 4 points – 48 respondents, 30 respondents – 3 points. This result allows us to conclude that the collective conflict level is low. Acceptance of risk. 35 respondents rated this parameter at 4 points, 37 at 3 points – this indicates that the company's staff is ready to take risks and believes that management will also take on such responsibility. The time spent on the implementation of the idea. Most respondents believe that the time spent on the idea is small. In our opinion, this indicates the absence of bureaucratization in the enterprise and the promotion of creativity.

Conclusion

Zammato's parameters intersect with those of Crowse and Exwell. We make brief conclusions on the results.
1. Conflict assessment is of positive importance – the majority of respondents consider the enterprise as a whole a non-conflict organization.
2. The parameter resistance to changes was rated quite high, which means that employees believe that there is no strong resistance to changes in the organization.
3. The majority of respondents believe that the transfer of responsibility is unusual for the team.

We assessed the socio-psychological climate in the work team of the enterprise using three intersecting approaches in order to increase the validity, reliability, and reliability of our study. In general, the situation is quite satisfactory, but some problems have been identified. Moreover, the problems, precisely, according to the key, in our opinion, parameters of the socio-psychological climate as the basis of organizational culture.

The team of the organization under study lacks cohesion, trust in management, support and recognition by management. Most employees rated their motivation low enough.

A specific area of manifestation of the socio-psychological climate in the team is the relationship between subordinates and the leader, leadership style. The leader must take care of the formation of a favorable socio-psychological climate in the team, because, otherwise, it will not be possible to use the group potential and develop an organizational culture.

In order for subordinates to feel the support and recognition of management, we recommend organizing effective feedback.

As we can see, the assessment of the socio-psychological climate (organizational climate) in the team is directly dependent on the organizational culture of the enterprise. A permanent assessment of its parameters allows, on the one hand, to correct the organizational culture in a timely manner, and on the other hand, to make management decisions during the process, since the presented assessment parameters can be considered as its main indicators.

References

Bodrova, M. I. (2019), "Human capital and social responsibility-sources for the formation of a highly effective organizational culture", Journal of Creative Economy, Vol. 13 No. 9, pp. 1635–1650.

Bodrova, M. I. (2019), " Development of organizational culture at enterprises in the Russian Federation – a driver of growth of the country's economy", Journal of Russian entrepreneurship, Vol. 20 No. 1, pp. 341–356.

Borscheva, A.V., Santalova, M.S., Soklakova, I.V., Surat, I.L. (2019) Innovation management in Russian business, Dashkov and K, Moscow.

Guskova, N.D. (2014) "The system of formation of organizational culture of an industrial enterprise based on management ethics", Journal of Leadership and management.,Vol. 1 No. 1, pp. 27–36.

Kachan, P.A. (2010) "The inverse effect of the motivational process in modern personnel management", Journal of Russian Entrepreneurship, Vol. 11 No. 8, pp. 76–80.

Kayl, I.I.,Zudina, E.V., Velikanov, V.V., Gaponenko, Y.V., Morozov, V.A. (2017). Innovational approach to management of human resourses of cluster entity. Contributions to Economics, (9783319454610),pp.189–195.

Morkovina, S.S., Popkova, E.G., Santalova, M.S., Konstantinov, A.V. (2014) "Development of methodological approaches to the efficiency analysis of territorial -industry cluster formation in the forest sector", Journal of Asian Social Science.,Vol. 10 No.23, pp. 85–94.

Nizovaya, I., Lesnikova, E., Nechaeva, S., Sadykova, K., Santalova, M.S. (2019). Crisis of the consumer basket in Russia. Lecture Notes in Networks and Systems, Vol. 57, pp. 852–861.

Rodionov, E.V., Santalova, M.S. (2012) Strategic management, Kvarta, Voronezh.

Santalova, M., Zemlyakov, D., Lesnikova, E., Fatyanova, I. (2017). Corporate culture of commercial organization as an effective management tool. Contributions to Economics, (9783319454610), PP. 101–110.

Santalova, M.S. and Petrov, D.S. (2014) "Evaluation of personnel and enterprises organization culture", Journal of Science and Society, No.2-1, P. 98.

Santalova, M.S., Borshcheva, A.V., Lesnikova, E.P., Nechaeva, S.N., Charykova, O.G. (2019). Information hindrances and communication barriers in project interactions. The Future of the Global Financial System: Downfall or Harmony, "Lecture Notes in Networks and Systems". Cham, pp. 273–281.

Santalova, M., Balahanova, D., Kuizheva, S., Lesnikova, E., Trunova, E. (2019). Effective tools for management of organization, Advances in Intelligent Systems and Computing, Vol. 726, pp. 537–545.

Santalova, M. S. (2006) "Social policy: do-juro and de facto", Journal of Man and labor, No. 6, pp. 30–31.

Santalova, M.S., Solomatina, E.D., Nikolaeva, Y.R. "How effective is the work with personnel in the trade organization", Journal of Science and Society, No. 2-1, pp. 136–138.

Elena A. Fomina, Julia V. Khodkovskaya, Ilvir I. Fazrakhmanov and Ekaterina E. Barkova

15 Harmonizing the Interests of the State and Business in Stakeholder Management

Introduction

The value concept of company management – the stockholder approach – is now the most common and more widely implemented in practice. The peculiarity of this approach is that all the company's activities are strictly subordinate to the goals of its owners, interpreted as increasing the cost of business, increasing the profitability of capital invested by the owners. The dominance of the stockholder approach is the main reason for the emergence of imbalances in the economy, structural deformities that are overcome through economic crises. One of the answers of modern science to the challenges of time was the stakeholder model of management.

Globalization processes in the context of the global economic crisis are approaching their completion, they are replaced by processes of localization through technological and technical modernization of the economy, strengthening the degree of state control over key sectors of the economy, and redistribution of budget resources in favor of the most priority projects. States place their national values, national interests, at the top of national policy priorities. In this regard, one of the main stakeholders becomes a state that takes into account the interests of the whole society as an embodied society. The authors presented the study, analyzed the best Russian and foreign practice of stakeholder management and proposed a mechanism for interacting the interests of the state and companies in the subsoil user and other nature-intensive business, taking into account the development of digital technologies.

Methodology

The methodological basis of this article was the research of leading Russian and foreign scientists, which set out the stockholder and stakeholder business models (Freeman., 2007), (Kayachev, 2019), etc. The nature of the relationship between stakeholders was studied on the basis of the classification proposed by P.-K. Mitchell,

---

Elena A. Fomina, Julia V. Khodkovskaya, Ilvir I. Fazrakhmanov, Ekaterina E. Barkova, Ufa branch of Financial University under the Government of the Russian Federation, Ufa, Russia

https://doi.org/10.1515/9783110654486-015

B.-R. Agle and D.-J. Wood in the model "Power, Legitimacy and Urgency" (Freeman, 2013).

Analyzing the content and essence of social competence (Mitchell., 1997), taking into account the principles of systemality, complexity, integration, accountability, dynamism, innovation, performance, etc., three basic business theories were investigated: corporate selfishness, corporate altruism, reasonable selfishness.

The study of the role and involvement of stakeholders in the activities of companies was carried out using a methodological and practical set of standards and manuals on social responsibility of the company (ISO 26000) and on interaction with interested agents (AA1000 SES). An analysis of the scientific approaches of various schools revealed the debating aspects of existing company management practices. This is due to the problems of adapting American (market, outsider) or European (related, insider) management models in the Russian business environment. Systematic analysis of the functioning and development of companies using the stakeholder model was carried out taking into account the identified objective difficulties of doing business in Russia: 1) the increasing limited natural resources in the most developed economic regions of the country, where natural sources of raw materials are depleted, the relative mass of potential labor is reduced, the area of free territory for the deployment of new productive forces is reduced; 2) due to the growth of oncoming traffic, the complexity of transport networks, the reduction of the capacity of the most stressed highways; 3) management problems are complicated when the state interests go against the departmental interests of a particular economic region and its management links. Objectively, these obstacles will increase in the future, and overcoming them by traditional methods, as practice has shown, is unproductive (Ostrovsky, 2008).

The formation of the mechanism of interaction between the state and business in the digital economy is carried out taking into account the stages of interaction between the company and stakeholders – stakeholders: 1) identification of groups of stakeholders; 2) identification of key interests of each stakeholder group; 3) identification of areas of possible interaction of stakeholders with the company; 4) assessing the effectiveness of stakeholder management. Since stakeholder management characterizes not only the efficiency of using resources that ensure the profitability of the business, but also the effectiveness of the company's interaction processes with interested agents, the need for the formation of such an interaction mechanism is updated, which, taking into account the interests of the state and business, is able to ensure the sustainable socio-economic development of society, adequately and flexibly respond to external and internal challenges. The proposed mechanism for harmonizing the interests of the state and business is built taking into account the specifics of subsoil user and other nature-intensive business, the measures applied and the effects of the state's regulatory impact on business in the digital economy.

Results

Forming the main goal of the company as satisfying the interests of all stakeholders, apologists of the stakeholder approach do not notice that various groups of stakeholders are in interaction, and quite often – in a very controversial one. At the same time, mitigating and even eliminating some of the contradictions is not an absolutely unsolvable task. So, for example, the expert community (Bain and others) has no doubt that the activities of companies to solve social issues of personnel really create value for both business and workers and their families. The developed standards and methodologies for the dissemination of the stakeholder model (AA1000 SES et al.) represent an attempt to create a theoretical basis for the functioning of companies in the conditions of effective interaction of a wide range of stakeholders. However, the share of such companies focusing on "value for all," and not only on "value for owners," is not yet large. The vast majority of companies see their surroundings as a warring side, attacking the interests of owners. The problems of almost antagonistic contradictions between the interests of capital owners and a wider range of persons, somehow falling into the orbit of the enterprise, interested in the consequences of its functioning, are known in the modern world. Already in the 90s of the last century, the main postulates of the stakeholder business model were formulated, which is designed to harmonize the interests of all involved in the economic turnover and related entities. Unfortunately, there are no well-defined algorithms, formalized methods of regulating business within the framework of the stakeholder model. In addition, the economic effects of implementing digital solutions in the entrepreneurial environment exacerbated the crisis not only in the development of technologies and industries, but also strengthened the multidirectional interests of business and society. A review of the practice of penetrating the principles of the stakeholder model into Russian management involves a number of reasons that impede the implementation of this approach. The analysis of causality in each case leads to the general conclusion that there is insufficient infrastructure in the broad sense of the word, which could allow the successful dissemination of the stakeholder model. The State, as the institution responsible for carrying out the functions of targeting and regulatory influence on socio-economic development, has a fundamental role in the creation of such infrastructure. The state, setting priorities related to security, communications, regulation of the business environment, is able to manage both technological and social development in the digital economy – the state's dual mission in implementing the stakeholder concept. On the one hand, the state, as a stakeholder, has the main interest in each company regarding tax revenues, on the other hand, the state is the only stakeholder who necessarily has relationships and interactions with all other stakeholders. The success of state tasks is directly related to how fully the interests of other stakeholders from the environment of each company are taken into account and realized. From this point of view, the state should be recognized as a key stakeholder, designed to create the required infrastructure and regulatory mechanism for the dissemination of stakeholder management that meets the needs of the digital business environment. Analyzing the foreign practice of the regulatory impact of the state on business, it should be clarified that the effect on the scale of the national economy from the introduction of digital technologies is characterized not only by the rate of GDP growth, but is also expressed as a qualitative change in the applied business models: accessibility and universality of use, flexibility in change management, embeddability in the concept of development. According to foreign experts, the faster the speed of diffusion of digitalization technologies in sectors of the economy, the higher the rate of economic growth is ensured.

Measures and effects of regulatory impact of the state on business are presented in Figure 15.1.

Figure 15.1: Measures and effects of the regulatory impact of the state on business.
Source: developed and compiled by the authors.

The established measures of state support for national projects in Russia do not take into account the consequences of technological development of the business, the impact on the ecosystem of implemented projects, interaction with stakeholders and the role in creating social value. Considering that building effective relationships with counterparties in stakeholder management is carried out taking into account types of economic activities, financial opportunities, needs, as well as values, recognition of the state's special role – the stakeholder leader is obvious. The proposed mechanism of interaction between the state and nature-intensive business is presented in Figure 15.2.

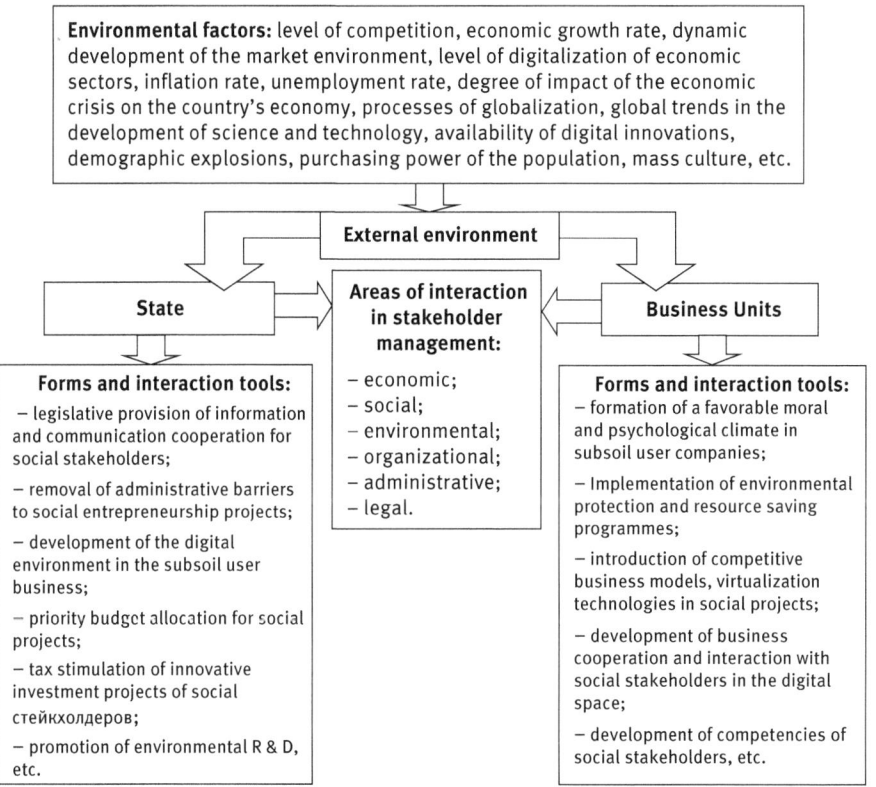

Figure 15.2: Mechanism of interaction between the state and business in stakeholder management. Source: developed and compiled by the authors.

In the mechanism of interaction between the state and business in the stakeholder management of subsoil use presented in Figure 15.2, the forms and tools of interaction are organized according to six perspectives (areas) of interaction, which consider aspects of economic, social, environmental, organizational, administrative and legal interaction. The mentioned prospects of interaction are very closely interconnected in subsoil users and other nature-intensive business, since exploration and extraction of raw materials and resources is often accompanied by man-made disasters, long-term negative impact on the environment. Environmental-intensive companies are characterized by the implementation of large-scale investment projects, the purpose of which is to obtain economic benefits from effective resource management (Fomina, 2018). Accordingly, assessing the efficiency of work and prospects of mining enterprises, first of all, it is important to take into account the dynamics of changes in mining and geological conditions during production, it is also necessary to assess the level of influence of these factors not only on the technical and economic indicators of production, but also on the environment. In the digital economy, a qualitatively built interaction with stakeholders contributes to increasing labor productivity and skills of labor resources, reducing prices, and facilitating access to information resources (Peskov., 2018). The key criteria in the nature-intensive business are limitations, optimality, flexibility, efficiency (Fomina, 2018). In addition, the criterion of economic efficiency of business projects is the most important for investors and managers, so obtaining the expected profit is controlled at all stages of investment development (Fomin, 2018) without taking into account "public value".

The issue of the need to solve such problems arises especially in the planned introduction of new production capacity, the reconstruction of existing capacities, choosing the order of re-equipment of organizations, when drawing up a production structure plan and in other situations. Errors in the design and operation of deposits associated with insufficiently studied geological anomalies, an accidental combination of natural and man-made factors cause accidents, explosions and fires that periodically occur in China in Poland, Russia, India, Canada, Germany and other countries (Ponomarenko., 2016). Effective capital investments in business reconstruction from the point of view of an individual company from the point of view of the industry may be ineffective from the point of view of society as a whole.

On the part of the state as a social stakeholder, regulatory impact should be highly effective. In this aspect, the formation of a single digital database of companies of subsoil users and other nature-intensive businesses, whose activities are associated with a negative impact on the environment, becomes important. Based on up-to-date information from a single database, the state can not only analyze the amount of damage, but also assess the potential environmental danger of each company and the entire industry. In addition, organizational decisions on the part of the social stakeholder-state should concern the development of social competence of business, moving to solve the problem of harmonizing the interests of business and society.

Conclusion

Therefore, the proposed mechanism of interaction between the interests of the state and companies in the subsoil user and other nature-intensive business allows, unlike existing approaches, to follow the concept of stakeholder management in the digital economy, and, taking into account the social competence of the business, to contribute to the formation of both financial and socially significant values. The coordination of the interests of society, business owners and the state should be influenced by the constructive, leading role of the state. The economic mechanism of legislative regulation of the use of subsoil in the digital economy needs a finer setup, providing a mutually beneficial or compromise solution to the issue. There can be no unilateral victory in this dispute: both the closure of the company and its uncontrolled and therefore destructive for nature activities carry huge costs. Measuring and estimating these costs goes beyond the cost concept of management and beyond the exclusive competence of capital owners. An analysis of international practice shows that the implementation of social investments by modern companies allows, using glass management tools, to create both corporate and social values. Measures of the state's regulatory influence on business are the foundation of the ongoing digitalization and should certainly be accompanied by both economic and social effects. The mechanism of interaction between the state and subsoil user and other nature-intensive business in stakeholder management differs from the model of value-oriented management in making management decisions aimed exclusively at achieving the interests of company owners. The proposed mechanism is based on the idea of the social value of the economic activity of companies: digitalization processes change not only business technologies, but also the multilateral interaction of economic entities. The development of digital space throughout the Russian Federation allows us to form a single information database on environmental monitoring of business impact on the environment, in terms of the size of harmful emissions of the largest companies as a subject of state regulation.

References

Birch, D. (1979), "The Job Generation Process", available at: https://ssrn.com/abstract=1510007 (accessed 27 May 2020).
Fomina, E.A., Khodkovskaya, Yu.V. (2018), "Resource management of business projects in the oil and gas industry," Economics and management: scientific and practical journal, No. 4 (142), p. 121–126.
Forecast of long-term socio-economic development of the Russian Federation for the period until 2030 (developed by the Ministry of Economic Development of Russia) http://www.consultant.ru/document/cons_doc_LAW_144190/ (circulation date: 08. 04.2020)
Freeman, R.E., Martin, K., Parmar, B. (2007), "Stakeholder Capitalism", Business Ethics Quarterly, Vol. 74, N 4, pp. 303–314.

Freeman, R.E., Moutchnik, A. (2013), "Stakeholder management and CSR: questions and answers", Forum on sustainable management, Vol. 1–2, URL:Nr.link.springer.com/article/10.1007/s00550-013-0266-3 (accessed on: 05. 05.2020).

Joek-Kovalska, I., Ponomarenko, T.V. and Marinina, O.A. (2018) "Problems of interaction with stakeholders in the implementation of long-term mining projects," Notes of the Mining Institute, Vol. 232, p. 428–437.

Kayachev, G.F., Loktionov, D.A. (2019), "Evolution of the value approach in company management," Leadership and management, Vol. 6, No. 4, p. 397–408.

Medovnikov, D.S., Rozmirovich, S.D., Hovhannisyan, TK (2016), "From Tech Success to national champions," p. 78.

Mitchell, R.-K., Agle, B.-R., Wood, D.-J. (1997), "Toward a theory of stakeholder identification and salience: defining the principle of who and what really counts", Academy of Management Review, Vol. 22, N 4, p. 853–886.

Mukhametdinova, A.R., Safarov, A.M., Magasumova, A.T., Khatmullina, A.T. (2012), "Assessment of the influence of the petrochemical complex enterprise on environmental objects," Georesource, No. 8 (50), p. 46–50.

Order of the Ministry of Economic Development of Russia from 27.06.2016 N 400 "On the priority project of the Ministry of Economic Development of Russia" Development of innovative clusters – leaders of world-class investment attractiveness, (access date: 17. 02.2020)

Ostrovsky, M.S., Rakhmanov, R.T. (2008), "Prerequisites for the development of an innovative process in the mining industry," Mining Information and Analytical Bulletin (scientific and technical journal), No. 2, p.34–36.

Peskov, D.R., Khodkovskaya, Yu.V., Sharafutdinov, R.B. "Digitalization of business processes in oil and gas companies," Eurasian Legal Journal, No. 9 (124), p. 438–444.

Ponomarenko, T.V. (2016), "Corporate social responsibility of the coal industry (the practice of Russian and European companies)," Notes of the Mining Institute, Vol. 222, p. 882–891.

Stakeholder Engagement and the Board: Integrating Best Governance Practices. Global Corporate Governance Forum. IFC. Focus 8, 2009. – p. 65 available at:https://www.ifc.org/wps/wcm/connect/bac56797-a3a7-4e24-90f6-efa9ab7363e0/FINAL%2BFocus8_5.pdf?MOD=AJPERES&CVID=jtCwtno (accessed 27 March 2020).

Lubov I. Vanchukhina, Tatyana B. Leybert, Elvira A. Khalikova, Ilnara R. Khanafieva and Giedrius Ciras

16 Methodological Approaches to Tender Procedures in Corporate Procurement Management in Companies

Introduction

In the context of the rapidly evolving procurement practices in the public and corporate sector of the economy of foreign countries and Russia remains one of the most urgent issue of open, transparent, and "healthy" competitive environment during the tendering and selection of suppliers. Efficiency of procurement policy is based on the well-organized vendor-selection process, taking into account the administrative procedures for procurement activities and best practices of its implementation in companies. In addition, when choosing a supplier it is necessary to consider two opposing factors that regulate the optimum choice of supplier during the tendering procedure – is the preservation and diversification. Preservation factor highlights the need for procurement with existing suppliers with experience in this market, a high level of staff qualification and obtained the status of a reliable supplier. Diversification factor is characterized by the application of the principles of alternative supplier selection in order to select the best conditions of delivery, better quality and lower cost Diversifying the suppliers, the customer creates a competitive environment, which has the right to choose and determine the best conditions of delivery, ensuring the effectiveness of procurement in the latter company (Forge, 2013).

Given the above, in the formation of policy and tender procedures for selecting suppliers for the delivery of material assets (services, works) in corporate procurement management system it is necessary to use modern methodical approaches. These approaches are based on the ranking and the diversification of suppliers, not only on the price criterion, but also take into account many important criteria that characterize the qualitative aspects of procurement in the company.

The authors presented the research analyzed the best Russian and foreign practice of procurement management in corporations regarding tender procedures and offer a multi-stage method of ranking and the subsequent selection of suppliers on the basis of price and non-price criteria.

Lubov I. Vanchukhina, Tatyana B. Leybert, Elvira A. Khalikova, Ilnara R. Khanafieva, Ufa State Petroleum Technological University, Ufa, Russia
Giedrius Ciras, Business technologies and entrepreneurship, VGTU Business management, Vilnius, Lithuania

https://doi.org/10.1515/9783110654486-016

Materials and Methods

Under the conditions of administering competitive procurements, Russian companies want to receive financial savings, but also provide deliveries with better material and technical resources, services and work. In this regard, the company-customer is subject to special requirements for participants in corporate procurement, and apply various methodological approaches to competitive selection of suppliers based on a combination of price and non-price criteria.

Establishment of criteria for the selection of suppliers or prequalification is carried out with the aim of:
1) reduce the risk of non-performance of the contract by the supplier
2) reduction of time for carrying out procurement procedures
3) ensuring increased publicity of procurement
4) ensuring the development of fair competition
5) the organization of continuous provision of the production process with material and technical resources, work and services
6) increase the efficiency of the procurement activities of the company

The main criterion for selecting a supplier is price. Non-price criteria also act as additional criteria: qualification of the participant, work experience, business reputation, financial condition and creditworthiness, and others.

The results of the analysis of the used Russian methods of prequalification of suppliers in the tender management system in theory are presented in Table 16.1.

Table 16.1: Analysis of the used methods of prequalification of suppliers in theory.

Method name / developer	Brief description of the method	Advantages of the technique	The disadvantages of the technique
Multi-criteria supplier selection based on hierarchy analysis / D.V. Kutuzov, E.P. Bykova (Kutuzov, Bykova, 2008)	This method allows the selection of suppliers on the basis of sequential pairwise comparison according to specified criteria. The mandatory criteria are the price of delivery, stability of delivery, the possibility of purchasing replacement products, the availability of a certified QMS, the presence of companies producing analogues, and the business reputation of the supplier	The ability to take into account a large number of criteria in the competitive selection of suppliers	Use of complex mathematical iteration

Table 16.1 (continued)

Method name / developer	Brief description of the method	Advantages of the technique	The disadvantages of the technique
Complex method of choosing the optimal supplier / E. Suchkova. (Suchkova, 2016)	This method allows you to select the optimal supplier for a variety of unequal criteria. Based on the selection of formalized and non-formalized selection criteria, and the rating of the selected criteria and the degree of their importance	Significance of criteria is based only on expert judgment	The method uses only non-price criteria for selecting suppliers. Applicable only to the supply of equipment
Methods of selection of suppliers based on the ranking of procurement by category and type of suppliers / Kozlova EV, VolynskyV.Yu. (Kozlova, Volynsky, 2015)	The proposed methodology and algorithm for the preliminary assessment of suppliers are based on the classification of purchases by the level of complexity and the assessment of the priority of suppliers. The final choice is made on the basis of the supplier's integral indicator of the priority by the method of additive convolution of the criteria and the weighting coefficient of the importance of the criterion by an expert.	Accounting for the category of procurement in the degree of complexity and the category of suppliers in terms of their reliability. Promotes greater transparency, more accurate and correct procurement strategies	

In the scientific literature and tendering practices in large companies, a wide range of techniques is used, which are based on a phased selection of suppliers, on a preliminary analysis of the priority of suppliers for non-price factors, on a preliminary multi-criteria selection based on hierarchy analysis.

The analysis of the methods of prequalification of suppliers in theory allows us to conclude that the proposed methodological approaches to grouping purchases by level of complexity and grouping suppliers by their degree of importance and reliability should be used in modern procurement management practices in Russian companies.

Given the established practice of conducting two-stage selection of suppliers in Russian companies, suppliers should be differentiated to ensure the reliability of the organization of competitive suppliers and the formation of a procurement strategy.

Results

It is proposed to conduct a two-stage tender based on price and non-price criteria. The proposed algorithm for the preliminary selection of the supplier in the tender management system in companies is presented in Figure 16.1.

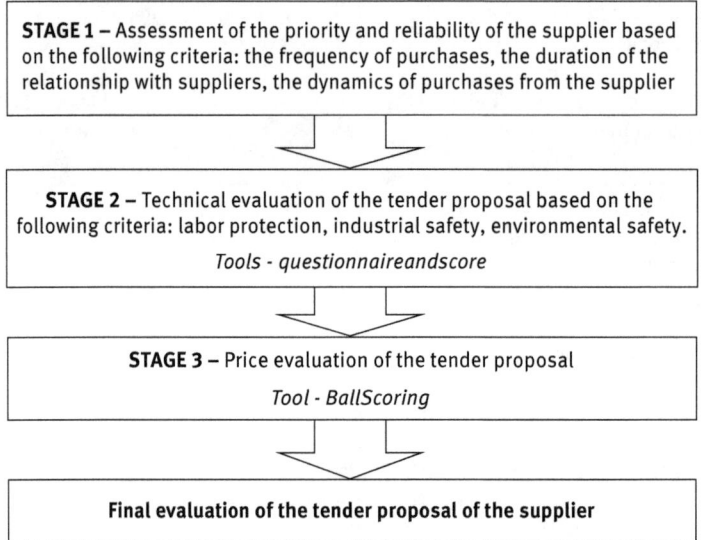

Figure 16.1: The proposed algorithm for the preliminary selection of the supplier in the tender management system in companies.

In accordance with the current rules for the administration of procurement procedures, a two-stage tender should be held with respect to the acquisition of capital facilities, research and development, supply of sophisticated equipment, and the provision of services (E. Suchkova, 2016).

At the first stage of the tender procedures, the priority and reliability of the supplier should be assessed based on the following criteria: reliability and reputation of the supplier, duration of partnerships, work experience and specialization, level of security.

At the second stage, a technical evaluation of the tender proposal is carried out, and at the third stage – a price evaluation of the tender offer, where the best supplier who offers the lowest price is selected.

Based on the technical and price estimates, the final evaluation of the tender proposal is displayed, that is, the place occupied by the tender proposal in the tender is determined by summing the technical and price estimates taking into account the significance factors of each of them.

Depending on the complexity of the work (services) and their threshold price, the ratio of the importance of technical and price estimates may vary. These ratios (coefficients) are presented in Table 16.2.

Table 16.2: Proposed Ratios.

Threshold price of the tender item, millions rub.	Technical evaluation (Kt)	Price estimate (Kp)
up to 15	4	6
from 15 to 30	5	5
over 30	6	4

The technical evaluation of supplier documentation is carried out on a 10-point scale, which takes into account factors such as occupational safety, industrial safety, environmental friendliness (environmental protection) and other parameters.

The technical evaluation of the tender proposal is based on an assessment of the factors determining the reliability of the potential supplier, his experience and ability to manufacture (deliver) the goods, as well as the conformity of the goods with the specified tender documentation with technical parameters.

The total assessment of the technical part of the tender offer $\sum T_n$ is determined by the following formula:

$$\sum T_n = \sum T_f \times K_z \qquad (1)$$

where

T_f – evaluation of the technical factor of the tender proposal on a 10-point scale
K_z – the weighting of technical factors

The results of the technical evaluation of each tender proposal shall be recorded and signed by experts.

Protocols are transmitted by experts to a senior expert who determines the arithmetic mean of each tender proposal by formula:

$$\bar{T} = \frac{\sum T_n}{N} \qquad (2)$$

where N – the number of experts who evaluate the technical part of the tender proposal.

The final evaluation of the technical part (Ct) of each registered applicant is determined by the formula:

$$C_t = \frac{\bar{T}}{Tmax} \qquad (3)$$

where

$\bar{T}_x$ – maximum estimate from arithmetic mean estimates of technical part.

The third stage is the price evaluation of the tender proposal (Cpi), which is carried out by the senior expert on the formula:

$$C_{pi} = \frac{Pmin}{P_i} \qquad (4)$$

where

P_{min} – minimum bid price;

Pi – price of the tender proposal of a particular applicant.

A price estimate characterizes the competitiveness of the tender offer and consists in determining the ratio of the minimum price offered in the tender proposals to the price of the tender proposal under consideration. In this case, the price is given taking into account all the costs of the supplier for the manufacture, delivery and other costs in accordance with the conditions specified in the tender documentation. If necessary, calculations are carried out to bring prices to a comparable form.

The final evaluation of the proposal is determined as the total of the two components of the technical evaluation (Ct) and price evaluation (Cp), multiplied by the corresponding coefficients of significance (Kt and Kp respectively) and is determined by:

$$C = Ct \times Kt + Cp \times Kp \qquad (5)$$

The final evaluation of the tender proposal determines the place of the tender proposal in the tender.

Discussion

Foreign practice shows that corporate procurement management is an area that is related to the supply chain of material and technical resources to meet the needs of the company's main production processes. The corporate procurement management system includes the following integral procedures and steps: development of a procurement strategy; information support for the procurement management system; organization of supplier relationships; and performance monitoring and risk assessment of procurement activities (M. Forge et al., 2013).

A large number of scientific publications focus on supply chain management, both in the context of the contract system and for the purpose of meeting the overall need for material resources (works and services) in corporations and large companies. Corporate procurement is not just a centralized supply of the necessary resources to support the company's core business processes, but a whole system that assesses the economic feasibility of using the material resources purchased in terms of product quality, technical and environmental safety, and estimating the cost of delivering and maintaining it throughout the life cycle. There are many models and approaches.

For example, public and corporate procurement management in the United States uses the technology of managing a single cycle of forecasting, placement, and use of contracts. The efficiency of procurement is significantly affected by strict regulation of procurement procedures, competitive bidding procedures using a single methodology and library of model contracts and a data bank for purchased goods. Centralized organization of procurement for public and corporate needs is intended to ensure information openness, transparency of procurement and healthy competition (P. Paulov et al., 2018).

In particular, in the scientific paper "Circular supply chain management: A definition and structured literature review," a group of authors presented a literary review of the application of supply chain management technology (CSCM), which is based on the purchase and supply of goods taking into account the assessment of their circular consumption and responsibility for the quality of the goods. This technology is also a system for managing complex supply chains of raw materials and components to the company, including the entire cycle of movement – from the purchase of raw materials and materials to the delivery of finished products on time to the end-user (M. Farooque et al., 2019).

Savchenko V.V. notes that SRM-sistem (Supplier Relationship Management System) (V. Savchenko, 2016) is used for optimal interaction with suppliers in foreign corporations. It is a set of consistent procedures used in the procurement system and aimed at the interaction of the company with potential suppliers. While in most Russian companies the traditional choice of supplier is based on the concept of obtaining procurement savings, in foreign practice procurement is the concept of a logistics system for basic business processes, based on the establishment of long-term partnerships and strong ties with suppliers.

SRM-sistem includes the following procedures:
- Segmentation of procurement by resources, suppliers and other parameters
- Ranking suppliers and setting key performance indicators
- Planning of procurement depending on the need for logistical resources, taking into account changes in the external and internal conditions of the company

Within SRM-sistem and Supplier Partnership, it is fundamental to develop a procurement strategy and total cost of ownership (J-S. Lin et al., 2011). Based on the results of procurement segmentation and supplier ranking, the total cost of ownership of the material and technical resource is determined, from the stage of procurement to the stage of its liquidation and write-off from production. The mandatory steps in the tender procedure are the pre-qualification of suppliers and the valuation of supply chain costs.

The group of authors (O.Karabak et al., 2019) in their scientific article gave an analysis of the main factors that have a significant impact on the reduction of the purchase price of the purchased goods during electronic auctions. The number of suppliers involved and the product category have a significant impact on the lower purchase price as the main factors. Therefore, in order to effectively manage corporate procurement, it is necessary to classify the purchased products into categories and to establish a procurement and supplier selection procedure for each of them.

Conclusion

Thus, the proposed supplier selection methodology, based on the multi-step evaluation of price and non-price criteria, taking into account the priority and reliability of suppliers, allows, unlike existing approaches, to follow the concept of three-factor development of the company in a strategic perspective based on environmental safety, industrial safety and social partnership. In this case, the supplier priority controlling should be carried out at the supplier prequalification stage.

References

Farooque M., Zhang A., Thürer M., T. Qu, D. (2019) "Huisingh Circular supply chain management: A definition and structured literature review" // Journal of Cleaner Production. – No.228. – P. 882–900.
Federal Law "On the procurement of goods, works, services by certain types of legal entities" (July 18, 2011) No. 223-FL // [Electronic resource]. – URL:http://base.garant.ru/ (appeal date 04/ 22/2018).
Forge, M.M. (2013). Procurement Guide: studies. Manual. Higher School of Economics, 695 p.
Karabağ O., Tan B. (2019) "An empirical analysis of the main drivers affecting the buyer surplus in E-auctions" // International journal of production research. – No.57(11). – P. 3435–3465.
Kozlova, E.V., Volynsky, V.Yu. (2015). Improving the process of preliminary assessment of suppliers of material resources at a machine-building enterprise. Economic analysis: theory and practice. 47–60.
Kutuzov, D.V., Bykova E.P. (2008). Application of the hierarchy analysis method to the choice of a supplier in the implementation of the "procurement" process of the quality management system. Caspian Journal: Management and High Technologies. 57–62.
Lin J-S.,Ou Jerry J.R. (2011) «A study on supply chain value-odded logistics based» // International Journal of Electronic Business Management. – Vol. 9. – No.1. – P. 58–69.
Monitoring the application of the Federal Law (July 18, 2011) No. 223-FL "On the procurement of goods, works, services by certain types of legal entities". [Electronic resource]. – URL:http://base.garant.ru/ (appeal date 04/ 22/2018).
Paulov P., Sotnikova V. (2018) "Experience of management of purchases in the USA" // Bulletin of Science and Practice. –Vol. 4.- No.11. – P. 401–404. – [Electronic resource] –// URL:http://www.bulletennauki.com.

Savchenko, V.V. (2016). Analysis of foreign experience in public and corporate procurement management.MID (Modernization.Innovations.Development). 166–169.

Suchkova, E.A. (2016). Comprehensive decision-making methods for choosing a supplier on the example of the purchase of dental equipment. Economics and Manage-ment: analysis of trends and development prospects. 249–254.

Conclusion: Future Perspectives of Industry 4.0

Industry 4.0 is a new modern reality, which is shown in the research results that are presented in this book. However, the strategic plans of development of Industry 4.0 were disturbed by an unexpected global socio-economic crisis caused by the COVID-19 pandemic in 2020. This led to a serious contradiction that is especially vivid in the sphere of management, economics, and law. On the one hand, Industry 4.0 – as a powerful mechanism of innovative development of society and economy – has a large potential of crisis management and requires the top-priority practical implementation.

On the other hand, in the conditions of the COVID-19 pandemic, all economic resources are used in the interests of healthcare and social protection. The importance of security provision, which is connected to cyber security in the context of Industry 4.0, has also grown. Transition to Industry 4.0 is connected to a high risk of emergence of a new crisis and unsolved problems of cyber security, which requires quick reaction with the international coordination of actions and cooperation.

The COVID-19 pandemic and crisis in 2020 have outlined new important directions of development of Industry 4.0: control of social interactions, digital monitoring of poverty, targeting of social support for population, and intellectual management in healthcare for increasing its flexibility in case of future epidemics and pandemics. This envisages creation of smart cities and regions and requires a thorough study and practical elaboration. In particular, it is necessary to look for a balance between general safety and excessive control from government. These issues have to be reconsidered from the positions of social ethics, regulatory support, institutionalization, and national economic policy – which should be done in further works.

Marina L. Alpidovskaya, Ludmila A. Karaseva, David I. Mamagulashvili,
Aleksei V. Bogoviz and Artem I. Krivtsov

List of Figures

Figure 1.1	RGC identification algorithm —— 7	
Figure 1.2	GRC selection algorithm for differentiation of state support measures —— 8	
Figure 3.1	Dynamics of the number of people employed in the industry of the Republic of Uzbekistan for 2000–2019 and forecast values for 2020–2023, (thousand person) —— 24	
Figure 4.1	Direct and total contribution of the tourism industry to world GDP, in billions of us dollars in 2006–2019 (WWTC, 2019; UNWTO, 2017) —— 33	
Figure 4.2	Transformation of economic relations of tourism industry subjects under the influence of digital technologies —— 39	
Figure 4.3	Model of joint creation of consumer value in the tourism industry Source: compiled by the authors —— 41	
Figure 6.1	Algorithm of project effectiveness monitoring organization —— 58	
Figure 6.2	Steps of project effectiveness monitoring process —— 58	
Figure 7.1	Algorithm for the generation of internal reports —— 65	
Figure 10.1	The relationship between types of accounting —— 89	
Figure 10.2	Relationship between types of accounting —— 89	
Figure 10.3	Relationship between accounting components —— 90	
Figure 10.4	Organizational structure of accounting —— 92	
Figure 15.1	Measures and effects of the regulatory impact of the state on business —— 136	
Figure 15.2	Mechanism of interaction between the state and business in stakeholder management —— 137	
Figure 16.1	The proposed algorithm for the preliminary selection of the supplier in the tender management system in companies —— 144	

List of Tables

Table 3.1	Dynamics of the main indicators of industry of the Republic of Uzbekistan for 2000–2019 —— 22	
Table 3.2	Industrial Innovation Indicators 2009–2018 —— 25	
Table 4.1	Digital technologies used in the tourism industry —— 35	
Table 8.1	Distribution of victims with an extreme situation by outcome, depending on the performance of the operation —— 70	
Table 9.1	Brief description of development of digital technologies in sectors of economy —— 79	
Table 9.2	Upgrades of components of DMS —— 81	
Table 10.1	Definition and criteria of differentiation of financial, managerial and production accounts —— 87	
Table 11.1	Prerequisites for the emergence and development of the sharing economy —— 97	
Table 11.2	Structure and content of socio-economic relations of the sharing economy —— 100	
Table 11.3	Risks of development of socio-economic relations of the sharing economy —— 103	
Table 12.1	Interpretation of expert evaluations concerning the impact of the fourth industrial revolution on the technological, managerial, economic, social, political, legal and environmental factors of world development using the SNW analysis method —— 108	
Table 14.1	Measurement of the organizational climate in the team (Coase) —— 124	
Table 14.2	Measurement of the socio-psychological (organizational) climate in the labor collective (Zammato and Krakovera) —— 125	
Table 14.3	Measurement of the parameters of the organizational climate of the company (Kouz) —— 127	
Table 14.4	Measurement of the parameters of the organizational climate in the work team (Exval) —— 128	
Table 14.5	Measurement of the socio-psychological climate in the labor collective according to Zammato and Krackover —— 128	
Table 16.1	Analysis of the used methods of prequalification of suppliers in theory —— 142	
Table 16.2	Proposed Ratios —— 145	

Index

Agricultural 77, 80

Balance 103, 110
Business 134, 135

Companies 146, 147
Corporate 146, 147
Critical 4, 66

Digital Technologies 63, 65
Digitalization 9, 32

Economic Relations 32, 34
Effectiveness 21, 26
Enterprises 4, 138

Finance 13, 16
Fourth Industrial Revolution 107, 108

Industry 4.0 107, 108
Innovative Activity 25, 27
Internal Control 32, 34

Loans 48

management 34, 42
Management Accounting System 62, 63
Modernization 78, 133
Monitoring 139, 146

Organizational Culture 123, 124

Pandemic 61
Political Economy 116

R&D 16, 25

Sharing Economy 95, 96
Socio-Economic Development 134
Stakeholder Management 136, 137
State 133, 134
Sustainable 117, 134

Technical 72, 73
Tender 141, 142
Tourism Industry 33, 34

Water Management Industry 80, 82
World Economy 107, 109